Automotive Technician Training: Practical Worksheets Level 1

Automotive practical worksheets for students at Level 1

This Level 1 student worksheets book contains tasks that help you develop practical skills and prepare you for assessment. The tasks also reinforce the automotive theory that you will learn online and in the classroom. Each worksheet covers individual topics in a step-by-step manner, detailing how to carry out all the most important tasks contained within the syllabus. Alongside each of these worksheets is a job card that can be filled in and used as evidence towards your qualification.

▶ Endorsed by the Institute of the Motor Industry for all their Level 1 automotive courses.
▶ Step-by-step guides to the practical tasks required for all Level 1 qualifications.
▶ Job sheets for students to complete and feedback sheets for assessors to complete.

Tom Denton is the leading UK automotive author with a teaching career spanning lecturer to head of automotive engineering in a large college. His range of automotive textbooks published since 1995 are bestsellers and led to his authoring of the Automotive Technician Training multimedia system that is in common use in the UK, USA and several other countries. Tom now works as the eLearning Development Manager for the Institute of the Motor Industry (IMI).

Automotive Technician Training

Practical Worksheets Level 1

Tom Denton

Routledge
Taylor & Francis Group

LONDON AND NEW YORK

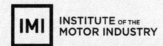

INSTITUTE OF THE
MOTOR INDUSTRY

First published 2015
by Routledge
2 Park Square, Milton Park, Abingdon, Oxon OX14 4RN

and by Routledge
711 Third Avenue, New York, NY 10017

Routledge is an imprint of the Taylor & Francis Group, an informa business

British Library Cataloguing-in-Publication Data
A catalogue record for this book is available from the British Library

Library of Congress Cataloging in Publication Data
A catalog record for this book has been requested

ISBN: 978-1-138-85236-5 (pbk)
ISBN: 978-1-315-72353-2 (ebk)

Typeset in Univers by
Servis Filmsetting Ltd, Stockport, Cheshire

Practical Worksheets – Level 1

Introduction

The purpose of this worksheets book is to provide a range of practical activities that will enable you to develop your abilities as a technician. The tasks are aligned with recognized vocational qualifications. However, there are far more tasks within this workbook than are required by the awarding body for the achievement of a Vocationally Recognized Qualification – because the more you practise, the more skills you will develop.

The worksheets are presented as three separate books at Level 1, Level 2 and Level 3 to follow the recognized qualifications. Within each level there are tasks for the major automotive areas: Engines, Chassis, Transmission and Electrical. The tasks range from component identification to removal and refit at Level 1 and 2, and diagnosis of complex system faults at Level 3.

A blank job card and assessor report are provided with each worksheet. This should be copied and then filled in alongside the task you are completing, including all relevant details regarding the vehicle, the fault and the rectification procedure where appropriate. You should write down a description of the work that you did to complete the task including any technical data that you sourced, any difficulties that you encountered and how you overcame them. If you had any health and safety issues to address, i.e. disposal of waste materials or clearing up spillages, this will help demonstrate your competence. By completing job cards thoroughly at this stage of your career as a technician, you will be well prepared for the time when you are required to complete job cards in the workplace. This can be very important, for example, if a warranty job card is not accurate then the manufacturer will not pay for the claim. An example of a completed job card is shown on page 7.

For teacher/lecturers, this workbook more than covers the requirements for Vocational Qualifications. Using the following tracking document you can note progress and also cross-reference the highlighted worksheets that directly relate to the awarding body required practical tasks.

Tracking

Engines (p.8)	12	24	35	45
1	13	25	36	46
2	14	26	37	47
3	15	27	38	48
4	16	Chassis (p.62)	Transmission (p.84)	49
5	17	28	39	50
6	18	29	40	51
7	19	30	41	52
8	20	31	42	53
9	21	32	43	54
10	22	33	Electrical (p.94)	55
11	23	34	44	56

Important notes about practical work

Safety

Working on vehicles is perfectly safe as long as you follow proper procedures. For all of the worksheets in this book you must therefore:

Comply with personal and environmental safety practices associated with clothing; eye protection; hand tools; power equipment; proper ventilation; and the handling, storage, and disposal of chemicals/materials in accordance with all appropriate safety and environmental regulations.

There are some specific recommendations below but you should also refer to the other textbooks or online resources for additional information.

Personal protective equipment (PPE), such as safety clothing, is very important to protect yourself. Some people think it clever or tough not to use protection. They are very sad and will die or be injured long before you! Some things are obvious, such as when holding a hot or sharp exhaust you would likely be burnt or cut! Other things such as breathing in brake dust, or working in a noisy area, do not produce immediately noticeable effects but could affect you later in life.

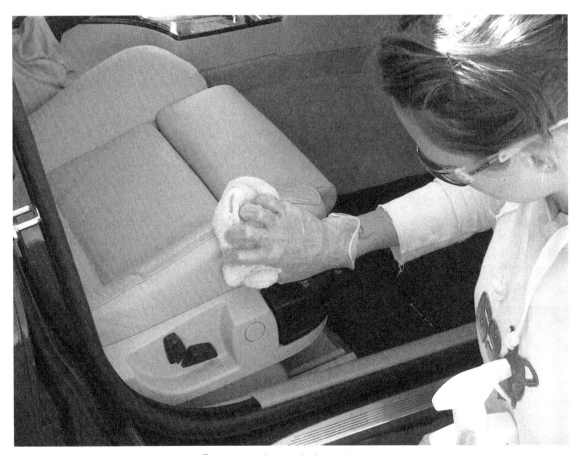

Eye protection and gloves in use

Fortunately the risks to workers are now quite well understood and we can protect ourselves before it is too late. In the following table, I have listed a number of items classed as PPE (personal protective equipment) together with suggested uses. You will see that the use of most items is plain common sense.

Equipment	Notes	Suggested or examples where used
Ear defenders	Must meet appropriate standards	When working in noisy areas or if using an air chisel
Face mask	For individual personal use only	Dusty conditions. When cleaning brakes or if preparing bodywork
High visibility clothing	Fluorescent colours such as yellow or orange	Working in traffic such as when on a breakdown
Leather apron	Should be replaced if it is holed or worn thin	When welding or working with very hot items
Leather gloves	Should be replaced when they become holed or worn thin	When welding or working with very hot items and also if handling sharp metalwork
Life jacket	Must meet current standards	Use when attending vehicle breakdowns on ferries!
Overalls	Should be kept clean and be flame proof if used for welding	These should be worn at all times to protect your clothes and skin. If you get too hot just wear shorts and a T-shirt underneath
Rubber or plastic apron	Replace if holed	Use if you do a lot of work with battery acid or with strong solvents
Rubber or plastic gloves	Replace if holed	Gloves must always be used when using degreasing equipment
Safety shoes or boots	Strong toe caps are recommended	Working in any workshop with heavy equipment
Safety goggles	Keep the lenses clean and prevent scratches	Always use goggles when grinding or when any risk of eye contamination. Cheap plastic goggles are much easier to come by than new eyes
Safety helmet	Must be to current standards	Under vehicle work in some cases
Welding goggles or welding mask	Check the goggles are suitable for the type of welding. Gas welding goggles are NOT good enough when arc welding	You should wear welding goggles or use a mask even if you are only assisting by holding something

Also, as well as your own protection you should always use a protection kit for the vehicle: floor mats, wing covers and seat covers for example.

Tools and equipment

To carry out any work you will need a standard toolkit and in some cases additional 'special' tools will be required. Make sure you have access to all necessary equipment before starting work. A few examples are mentioned below but you should also refer to the other textbooks or online resources for additional information.

Using hand tools is something you will learn by experience, but an important first step is to understand the purpose of the common types. This section therefore starts by listing some of the more popular tools, with examples of their use, and ends with some general advice and instructions.

Toolkit

Practise until you understand the use and purpose of the following tools when working on vehicles.

Hand tool	Example uses and/or notes
Adjustable spanner (wrench)	An ideal stand by tool and useful for holding one end of a nut and bolt.
Open-ended spanner	Use for nuts and bolts where access is limited or a ring spanner can't be used.
Ring spanner	The best tool for holding hexagon bolts or nuts. If fitted correctly it will not slip and damage both you and the bolt head.
Torque wrench	Essential for correct tightening of fixings. The wrench can be set in most cases to 'click' when the required torque has been reached. Many fitters think it is clever not to use a torque wrench. Good technicians realize the benefits.
Socket wrench	Often contain a ratchet to make operation far easier.
Hexagon socket spanner	Sockets are ideal for many jobs where a spanner can't be used. In many cases a socket is quicker and easier than a spanner. Extensions and swivel joints are also available to help reach that awkward bolt.
Air wrench	These are often referred to as wheel guns. Air-driven tools are great for speeding up your work but it is easy to damage components because an air wrench is very powerful. Only special, extra strong, high-quality sockets should be used.
Blade (engineer's) screwdriver	Simple common screw heads. Use the correct size!
Pozidrive, Philips and cross-head screwdrivers	Better grip is possible particularly with the Pozidrive but learn not to confuse the two very similar types. The wrong type will slip and damage will occur.
Torx®	Similar to a hexagon tool like an Allen key but with further flutes cut in the side. It can transmit good torque.
Special purpose wrenches	Many different types are available. As an example mole grips are very useful tools as they hold like pliers but can lock in position.
Pliers	These are used for gripping and pulling or bending. They are available in a wide variety of sizes. These range from snipe nose, for electrical work, to engineers pliers for larger jobs such as fitting split pins.
Levers	Used to apply a very large force to a small area. If you remember this you will realize how, if incorrectly applied, it is easy to damage a component.
Hammer	Anybody can hit something with a hammer, but exactly how hard and where is a great skill to learn!

General advice and instructions for the use of hand tools (taken from information provided by Snap-on):

▶ Only use a tool for its intended purpose
▶ Always use the correct size tool for the job you are doing
▶ Pull a spanner or wrench rather than pushing whenever possible
▶ Do not use a file or similar without a handle
▶ Keep all tools clean and replace them in a suitable box or cabinet
▶ Do not use a screwdriver as a pry bar
▶ Look after your tools and they will look after you!

Information

Before starting work you should always make sure you have the correct information to hand. This can be in the form of a workshop manual or a computer-based source.

The worksheets in this book are a general guide so make sure the correct information, procedures and data for the particular vehicle you are working on are available before you start work.

 Technical data

 Timing belts

 Timing chains

 Timing gears

 Service indicator reset procedures

 Key programming

 Manufacturers' service schedules

 Service illustrations

 Repair times

 Wheel alignment

 Diagnostic trouble codes

 Tyre sizes and pressures

 Known fixes and bulletins

 Engine management component testing

 Engine management pin data

 Engine management trouble shooter

 Airbags

 Anti-lock brake systems

 Air conditioning

 Electrical component locations

 Wiring diagrams

 Guided diagnostics

 Tyre pressure monitoring system

 Electric parking brake

 Battery disconnection and reconnection procedures

Autodata online information

Job card: example

Technician/Learner name & date	Make and model	VIN no.	Reg. no.	Job/task no.
John Doe	Ford Mondeo	1M8GDM9A_KP042788	ABC 123	100

Customer's instructions / Vehicle fault	Mileage	67834

Carry out minor service – change oil and filter.

Inspect brakes.

Check for rattle underneath when accelerating.

Work carried out and recommendations (include PPE & special precautions taken)

PPE worn – boot, gloves and overall, VPE – wing covers, floor mats and steering wheel cover. Followed service checklist, full under bonnet check of belts, for oil, fuel and coolant leaks. Drained oil and replaced filter, torque new filter to 15Nm as per manufacturer's instructions, filled engine with 6 litres of engine oil. Old engine disposed of in oil drum and filter placed in specific filter bin.

Full under vehicle check, hoses, brakes pipes, all steering and suspension components – all ok. Found detached exhaust mounting – this would cause the knock the customer complained of, replaced mounting.

Checked brakes, pads and discs ok 30% wear on pads.

Requires 2 front tyres, have notified customer but they will go to Kwikfit.

During the service a quantity of oil was spilled on the floor, I covered the spillage with granules and left them to soak the oil up. I then cleaned the granules up and disposed of them in the correct bin. Finally I mopped the floor to ensure that it was properly clean.

Parts and labour	Price
3 hours @ 22.50 per hour	£67.50
Oil	£18
Oil Filter	£6.80
Exhaust mounting	£14
Total	£106.30

Data and specifications used (include the actual figures)

Oil filter torque – 15Nm, Wheel nut torque – 160Nm, 6.0 litres of engine oil

Assessor report: example

	Assessment outcome	Passed (tick ✓)
1	The learner worked safely and minimised risks to themselves and others	✓
2	The learner correctly selected and used appropriate technical information	✓
3	The learner correctly selected and used appropriate tools and equipment	✓
4	The learner correctly carried out the task required using suitable methods and testing procedures	✓
5	The learner correctly recorded information and made suitable recommendations	✓

Assessor name (print)	Tick	Written feedback (with reference to assessment criteria) must be given when a learner is referred
PASS: I confirm that the learner's work was to an acceptable standard and met the assessment criteria of the unit	✓	*Candidate worked in a very organised manner.* *Work area was kept clean and tidy throughout, tools returned to toolbox once used and oil and filter disposed of correctly.* *Good communication regarding further work requirements found during the task.* *Assessment criteria met, well done.*
REFER: The work carried out did not achieve the standards specified by the assessment criteria		

Assessor Name (Print)	Assessor PIN/Ref.	Date
Jane Jones	1234	29/02/17

The section below is only to be completed by the learner once the assessor decision has been made and feedback given to learner			
I confirm that the work carried out was my own, and that I received feedback from the Assessor	Learner name (Print)	Learner signature	Date
	John Doe	J Doe	29/02/17

Engines

Worksheet 1: Routine vehicle maintenance inspections/service

Procedure

▶ General visual inspection – listen for abnormal noises.

▶ Visual/oil level and condition. Replace oil and filter at specified intervals.

▶ Visual/inspections for oil leakage.

▶ Visual/inspection of exhaust smoke – at idle speed at mid-engine speed (3000 rpm) on overrun during road test.

▶ Engine oil pressure test – check warning light operates (attach a pressure gauge and adapter and tachometer) – pressure at idle speed – stabilized pressure – stabilized pressure engine rpm.

▶ Crankcase ventilation system – check condition of hoses (visual and remove hoses and valves) – orifice to inlet manifold clear – air cleaner condition – control valve condition.

▶ Check coolant level and specific gravity.

▶ Check brake operation and pad/disc condition – record thickness.

▶ Check tyre tread depths.

▶ Torque wheel nuts – include torque figures and calibration date of torque wrench.

▶ Additional items – see manufacturer's schedule.

Job card

Technician/learner name & date	Make and model	VIN no.	Reg. no.	Job/task no.

Customer's instructions/vehicle fault		Mileage	

Work carried out and recommendations (include PPE & special precautions taken)

	Price
Parts and labour	
Total	

Data and specifications used (include the actual figures)

Assessor report

	Assessment outcome	Passed (tick ✓)
1	The learner worked safely and minimised risks to themselves and others	
2	The learner correctly selected and used appropriate technical information	
3	The learner correctly selected and used appropriate tools and equipment	
4	The learner correctly carried out the task required using suitable methods and testing procedures	
5	The learner correctly recorded information and made suitable recommendations	

	Tick	Written feedback (with reference to assessment criteria) must be given when a learner is referred
Pass: I confirm that the learner's work was to an acceptable standard and met the assessment criteria of the unit		
Refer: The work carried out did not achieve the standards specified by the assessment criteria		

Assessor name (print)	Assessor PIN/ref.	Date

Section below only to be completed by the learner once the assessor decision has been made and feedback given

I confirm that the work carried out was my own, and that I received feedback from the Assessor	Learner name (print)	Learner signature	Date

Worksheet 2: Perform visual and aural checks on running engine

Procedure

▶ Lift bonnet – general impression – check oil and coolant levels. Start engine from cold and carry out checks as engine warms up.

▶ Check cooling system operation as engine warms up. Check top hose remains cool until engine is hot and then top hose should suddenly heat up as the thermostat opens. Electric fan operation follows when the engine temperature rises higher – keep hands clear of fan.

▶ Check fast idle speed (engine cold) and idle speed (engine hot).

▶ Look at the colour of the exhaust smoke, blip throttle and watch for changes.

▶ Increase engine speed to about 3000 rpm and hold for 20–30 seconds – watch exhaust smoke. Release throttle to allow engine to idle – watch exhaust smoke on deceleration.

▶ Look for oil leaks and oil flow from areas of leakage.

▶ Look at engine security when blipping throttle – look for excessive movement. Listen for exhaust knocking on body/chassis.

▶ Listen for abnormal noises – belt slipping squeal, scream/squeal from loose or leaking exhaust manifold or flange.

▶ Listen for engine mechanical sounds – light knocking/tapping from valve gear or pistons . . . Heavy knocking from crankshaft big end bearings . . . Rumble from crankshaft main bearings . . .

Job card

Technician/learner name & date	Make and model	VIN no.	Reg. no.	Job/task no.

Customer's instructions/vehicle fault		Mileage		

Work carried out and recommendations (include PPE & special precautions taken)

Parts and labour	Price
Total	

Data and specifications used (include the actual figures)

Assessor report

	Assessment outcome	Passed (tick ✓)
1	The learner worked safely and minimised risks to themselves and others	
2	The learner correctly selected and used appropriate technical information	
3	The learner correctly selected and used appropriate tools and equipment	
4	The learner correctly carried out the task required using suitable methods and testing procedures	
5	The learner correctly recorded information and made suitable recommendations	

	Tick	Written feedback (with reference to assessment criteria) must be given when a learner is referred
Pass: I confirm that the learner's work was to an acceptable standard and met the assessment criteria of the unit		
Refer: The work carried out did not achieve the standards specified by the assessment criteria		

Assessor name (print)	Assessor PIN/ref.	Date

Section below only to be completed by the learner once the assessor decision has been made and feedback given			
I confirm that the work carried out was my own, and that I received feedback from the Assessor	Learner name (print)	Learner signature	Date

Worksheet 3: Inspect chain drive mechanisms

Procedure

▶ Strip to access chain drive mechanism. Look for general condition, tension, lubrication and damage to covers.

▶ Check chain tension and condition of chain self-tensioner device. Check chain wear or stretch and 'fit' on the sprockets.

▶ Turn engine to align the timing marks and remove the chain and sprockets.

▶ Check chain wear by observing the bend when the chain is held on its side. Check all rollers for tightness or seizure.

▶ Check the condition of the teeth on the sprockets.

▶ Check sprocket to hub location and securing holes and devices.

▶ Reassemble chain and sprockets keeping the timing marks aligned. Refit and set tensioner. Turn engine at least two full revolutions and recheck timing and chain tension. When engine is running listen for abnormal noises and correct operation of the engine.

Job card

Technician/learner name & date	Make and model	VIN no.		Reg. no.	Job/task no.

Customer's instructions/vehicle fault		Mileage	

Work carried out and recommendations (include PPE & special precautions taken)

Parts and labour		Price
Total		

Data and specifications used (include the actual figures)

Assessor report

	Assessment outcome	Passed (tick ✓)
1	The learner worked safely and minimised risks to themselves and others	
2	The learner correctly selected and used appropriate technical information	
3	The learner correctly selected and used appropriate tools and equipment	
4	The learner correctly carried out the task required using suitable methods and testing procedures	
5	The learner correctly recorded information and made suitable recommendations	

	Tick	Written feedback (with reference to assessment criteria) must be given when a learner is referred
Pass: I confirm that the learner's work was to an acceptable standard and met the assessment criteria of the unit		
Refer: The work carried out did not achieve the standards specified by the assessment criteria		

Assessor name (print)	Assessor PIN/ref.	Date

Section below only to be completed by the learner once the assessor decision has been made and feedback given			
I confirm that the work carried out was my own, and that I received feedback from the Assessor	Learner name (print)	Learner signature	Date

Worksheet 4: Remove and replace air cleaners, carburettors, inlet manifolds and early fuel evaporation system components

Procedure

▶ Disconnect the battery earth lead. Undo the air cleaner and ducting securing screws/clips. Lift off air cleaner and ducting. Label and disconnect the throttle, choke and electrical cables to the carburettor and the early fuel evaporation heater grid. Disconnect and cap or plug the fuel feed and return pipes/hoses.

▶ Undo the carburettor to inlet manifold securing nuts and remove. Lift off the carburettor from the inlet manifold. Inspect all gaskets and the heat insulation block or the early fuel evaporative heater grid plate. Label and disconnect the vacuum connections on the inlet manifold. For water-heated inlet manifolds, drain the cooling system and disconnect the coolant pipes/hoses on the inlet manifold.

▶ Undo the inlet manifold to cylinder head nuts/bolts and remove. Lift off the manifold. Inspect the gaskets for signs of leakage. Cleaning all mating faces – cylinder head, manifold, carburettor and heat shield and check for true (flatness). Clean bolt or stud threads if necessary.

▶ Reassemble using new gaskets and either a soft setting sealant, where specified, or apply a smear of grease to the gaskets, particularly the carburettor and heat insulator block. Tighten all nuts/bolts to specified torque and in sequence (centre outward).

▶ Refit fuel pipes/hoses, coolant hoses and top up cooling system, refit throttle, choke and electrical cables. Check throttle cable adjustment gives full throttle (throttle plate vertical in carburettor) and choke control is fully off and can be fully applied.

▶ Run engine and check for correct operation. Check coolant level, bleed and top up if necessary. Check in-car heater operation. Check and adjust idle and fast idle speeds and mixture strength (CO). On vehicles with an early fuel evaporation system the heater grid plate replaces the heat insulation block. Lift off from the inlet manifold carefully to avoid damage. The electrical feed is controlled by a temperature switch in the engine coolant and by a relay. Check with the vehicle manufacturer's data for the location, testing, removal and reassembly procedures.

Job card

Technician/learner name & date	Make and model	VIN no.		Reg. no.	Job/task no.

Customer's instructions/vehicle fault		Mileage		

Work carried out and recommendations (include PPE & special precautions taken)

Parts and labour	Price
Total	

Data and specifications used (include the actual figures)

Assessor report

	Assessment outcome	Passed (tick ✓)
1	The learner worked safely and minimised risks to themselves and others	
2	The learner correctly selected and used appropriate technical information	
3	The learner correctly selected and used appropriate tools and equipment	
4	The learner correctly carried out the task required using suitable methods and testing procedures	
5	The learner correctly recorded information and made suitable recommendations	

	Tick	Written feedback (with reference to assessment criteria) must be given when a learner is referred
Pass: I confirm that the learner's work was to an acceptable standard and met the assessment criteria of the unit		
Refer: The work carried out did not achieve the standards specified by the assessment criteria		

Assessor name (print)	Assessor PIN/ref.	Date

Section below only to be completed by the learner once the assessor decision has been made and feedback given			
I confirm that the work carried out was my own, and that I received feedback from the Assessor	Learner name (print)	Learner signature	Date

Worksheet 5: Inspect and replace sump, covers, gaskets and seals

Procedure

▶ Disconnect battery ground cable.

▶ Paper gaskets – remove cover/housing and clean old gasket material and sealant from both gasket faces. Check faces for flatness, rectify as necessary. Fit new gasket with sealant, if specified, refit bolts and tighten to specified torque.

▶ Rubber gaskets – remove cover/housing and clean old gasket material from both faces. Check gasket faces for flatness, rectify as necessary. Fit new gasket without sealant to clean and dry faces, refit bolts and tighten until gasket is pinched. Do not over tighten.

▶ Cork gaskets – remove cover/sump and clean old gasket material and sealant from both gasket faces. Check faces for flatness, rectify as necessary. Fit new gasket with sealant, if specified, refit bolts and tighten to specified torque and in sequence.

▶ Formed in place gaskets (tube applied) – remove cover/sump and clean old gasket sealant from both faces. Check faces for flatness, rectify as necessary. Apply sealant bead gaskets as per manufacturer's instructions. Fit and torque in sequence within 3 minutes.

▶ Oil seals – these are used with gaskets for sealing around bearing caps and oil moulded rubber pans. Fit with recommended sealants or dry and then fit other types of gaskets and sump. Fit and torque bolts in sequence. Check seals do not squeeze out.

▶ Oil seals (lip) – remove pulley from housing and inspect seal land. Use a special tool to extract the seal. Press in new seal to shoulder or specified depth. Lubricate (in block) seal with clean engine oil. Clean and lubricate seal land, refit pulley and torque bolt.

Job card

Technician/learner name & date	Make and model	VIN no.	Reg. no.	Job/task no.

Customer's instructions/vehicle fault		Mileage	

Work carried out and recommendations (include PPE & special precautions taken)

Parts and labour	Price
Total	

Data and specifications used (include the actual figures)

Assessor report

	Assessment outcome	Passed (tick ✓)
1	The learner worked safely and minimised risks to themselves and others	
2	The learner correctly selected and used appropriate technical information	
3	The learner correctly selected and used appropriate tools and equipment	
4	The learner correctly carried out the task required using suitable methods and testing procedures	
5	The learner correctly recorded information and made suitable recommendations	

	Tick	Written feedback (with reference to assessment criteria) must be given when a learner is referred
Pass: I confirm that the learner's work was to an acceptable standard and met the assessment criteria of the unit		
Refer: The work carried out did not achieve the standards specified by the assessment criteria		

Assessor name (print)	Assessor PIN/ref.	Date

Section below only to be completed by the learner once the assessor decision has been made and feedback given			
I confirm that the work carried out was my own, and that I received feedback from the Assessor	Learner name (print)	Learner signature	Date

Worksheet 6: Inspect oil galleries and drillings in engine components

Procedure

▶ Strip the engine to access all galleries and oil drillings in the block. Identify galleries and drillings. Remove plugs in ends of main gallery to ensure full check is completed. Check the manufacturer's specifications for any special instructions of details.

▶ Use a cleaning tank with a low-pressure jet to provide a flow of cleaning solvent through the galleries and drillings.

▶ Flush through and check main gallery. Refit plugs with thread sealant. Flush through all other drillings, blank off each in turn to direct flow to other drillings. Check feeds to main bearings, camshaft, cylinder head and other components.

▶ Flush through the crankshaft to check main bearing to big end feeds. Flush through connecting rods to check for oil throw to cylinder walls and underside of piston crowns.

▶ Flush through all other hollow oil feed components.

▶ If slugging has occurred it may be possible to clear without stripping the engine. Use a special flushing oil that is used in the engine for a short period of time at a high idle speed. Follow the oil manufacturer's instructions. Renew the oil and filter.

▶ Where slugging is found, recommend frequent oil and filter changes in the future.

Job card

Technician/learner name & date	Make and model	VIN no.		Reg. no.	Job/task no.

Customer's instructions/vehicle fault	Mileage	

Work carried out and recommendations (include PPE & special precautions taken)

Parts and labour	Price
Total	

Data and specifications used (include the actual figures)

Assessor report

	Assessment outcome	Passed (tick ✓)
1	The learner worked safely and minimised risks to themselves and others	
2	The learner correctly selected and used appropriate technical information	
3	The learner correctly selected and used appropriate tools and equipment	
4	The learner correctly carried out the task required using suitable methods and testing procedures	
5	The learner correctly recorded information and made suitable recommendations	

	Tick	Written feedback (with reference to assessment criteria) must be given when a learner is referred
Pass: I confirm that the learner's work was to an acceptable standard and met the assessment criteria of the unit		
Refer: The work carried out did not achieve the standards specified by the assessment criteria		

Assessor name (print)	Assessor PIN/ref.	Date

Section below only to be completed by the learner once the assessor decision has been made and feedback given			
I confirm that the work carried out was my own, and that I received feedback from the Assessor	Learner name (print)	Learner signature	Date

Worksheet 7: Remove and replace oil pumps and drive mechanisms – time oil pump and distributor shaft

Procedure

▶ *Oil pump drives distributor.* If the oil pump and the distributor share the same drive from the camshaft, position the engine so that No. 1 cylinder is at TDC on the compression stroke. Remove the distributor cap and mark or record the position of the distributor and rotor arm. Remove the distributor and record the position of the drive dogs (large 'D' position). Replaced oil pump must return to this position when reassembled.

▶ On the underside of the engine disconnect and remove the oil pickup pipe and strainer. Undo and remove the oil pump securing bolts and remove the pump. Clean old gasket from mating faces. Strip and inspect pump. Refit pump with new gasket or 'O' ring and sealant *only* if specified. Turn drive gear to the timing position and ease home on the camshaft gear. Check the 'D' drives dog position in the distributor-housing bore. Correct if necessary by pulling back and turning the gear appropriately. When the distributor drive is correctly positioned, fit and tighten the securing bolts to the specified torque.

▶ Fit the pickup pipe and strainer using new seals and tighten securing bolts. Refit the distributor and set ignition timing. Finally adjust timing when the engine is running.

▶ *Oil pump not timed.* If the oil pump is driven from the crankshaft or camshaft or an auxiliary shaft independently of the ignition system, the pump drive mechanism need not be timed. Remove pickup pipe and strainer if necessary. Undo oil pump securing bolts and remove. Clean old gaskets or 'O' rings from pump and engine. Strip and inspect oil pump. Prime new pump with clean engine oil. Fit new gaskets or 'O' rings and sealant *only* if specified. Turn drive to locate with drive gear, dog or 'D' drive and fit pump. Fit and tighten securing bolts to specified torque. Reassemble engine and top up with engine oil. Before fitting new oil filter, *prime* the pump with clean engine oil through the oil feed hole to the oil filter. Fit filter. Start and run engine – if oil warning light does not go out in normal time, stop the engine and investigate. Look under the engine for oil leaks as soon as the oil warning light goes out.

Job card

Technician/learner name & date	Make and model	VIN no.	Reg. no.	Job/task no.

Customer's instructions/vehicle fault	Mileage	

Work carried out and recommendations (include PPE & special precautions taken)

Parts and labour	Price
Total	

Data and specifications used (include the actual figures)

Assessor report

	Assessment outcome	Passed (tick ✓)
1	The learner worked safely and minimised risks to themselves and others	
2	The learner correctly selected and used appropriate technical information	
3	The learner correctly selected and used appropriate tools and equipment	
4	The learner correctly carried out the task required using suitable methods and testing procedures	
5	The learner correctly recorded information and made suitable recommendations	

	Tick	Written feedback (with reference to assessment criteria) must be given when a learner is referred
Pass: I confirm that the learner's work was to an acceptable standard and met the assessment criteria of the unit		
Refer: The work carried out did not achieve the standards specified by the assessment criteria		

Assessor name (print)	Assessor PIN/ref.	Date

Section below only to be completed by the learner once the assessor decision has been made and feedback given			
I confirm that the work carried out was my own, and that I received feedback from the Assessor	Learner name (print)	Learner signature	Date

Worksheet 8: Remove and replace camshaft and bearings (OHV) and auxiliary shafts and bearings

Procedure

▶ Strip engine to access camshaft, remove push rods and cam followers, timing chain and camshaft sprocket and sump. (Remove distributor and oil pump drive gears if necessary.)

▶ Check camshaft end float with feeler gauges or dial test indicator (DTI). Undo securing bolts on camshaft retaining plate, remove plate and inspect for wear, particularly if end float is excessive.

▶ Pull camshaft forward out of bearings – use care to avoid damage to the bearings. Inspect bearing journals, cam lobes and gear teeth on the camshaft. Inspect the bearings (bushes) for condition.

▶ To remove the bearings, use the special tools and follow the manufacturer's instructions. It may be necessary to remove a plug to gain access to the rear bearing.

▶ Fit a new set of bearings with the special tools and ream or scrape if specified.

▶ Lubricate all bearings with clean engine oil. Feed the camshaft carefully through the bearings.

▶ Fit the retaining plate and measure the end float, compare with the specification – replace if the end float is not within tolerance. Torque bolts.

▶ Fit the drive chain and sprockets and check that the timing marks are correctly aligned. Turn engine for two or three complete revolutions and recheck timing and chain tension.

▶ Rebuild the engine – run and check for abnormal noises and correct operation. Road test and recheck under bonnet for abnormal noises and oil leaks etc.

▶ The procedure for auxiliary and balance shafts is similar to the above procedure for camshafts.

Job card

Technician/learner name & date	Make and model	VIN no.		Reg. no.	Job/task no.

Customer's instructions/vehicle fault		Mileage			

Work carried out and recommendations (include PPE & special precautions taken)

Parts and labour	Price
Total	

Data and specifications used (include the actual figures)

Assessor report

	Assessment outcome	Passed (tick ✓)
1	The learner worked safely and minimised risks to themselves and others	
2	The learner correctly selected and used appropriate technical information	
3	The learner correctly selected and used appropriate tools and equipment	
4	The learner correctly carried out the task required using suitable methods and testing procedures	
5	The learner correctly recorded information and made suitable recommendations	

	Tick	Written feedback (with reference to assessment criteria) must be given when a learner is referred
Pass: I confirm that the learner's work was to an acceptable standard and met the assessment criteria of the unit		
Refer: The work carried out did not achieve the standards specified by the assessment criteria		

Assessor name (print)	Assessor PIN/ref.	Date

Section below only to be completed by the learner once the assessor decision has been made and feedback given			
I confirm that the work carried out was my own, and that I received feedback from the Assessor	Learner name (print)	Learner signature	Date

Worksheet 9: Inspect oil sealing devices, gaskets, seals, etc.

Procedure

▶ Look closely at the 'fit' of lip-type seals in their housings for correct position, distortion, leakage bypassing the outside of the seal against the housing or any other possible routes.

▶ Look closely at the seal lips for burning, hardening or cracks. Look closely at the seal land on the shaft for undercutting, underscoring or other abnormal wear. Always lubricate and protect seal lips before and during the fitting of the seals over shafts.

▶ Check the tightness/torque of housings and covers with oil leakage from gaskets. Check the mating surfaces for flatness, distortion around boltholes, etc. with straight edge and feeler gauges or a surface table and engineer's blue.

▶ Check pressed steel covers with cork or rubber gaskets for over tightening and distortion of the gaskets. Flatten stretched boltholes with a hammer and flat surface.

▶ Check through bolts into engine crankcase and crankshaft flanges that the correct type of thread sealant has been used and correctly applied.

▶ Check other sealing devices for correct positioning, distortion incorrect sealant used, etc. Follow closely manufacturer's instructions/specifications for all sealing applications.

Job card

Technician/learner name & date	Make and model	VIN no.		Reg. no.	Job/task no.

Customer's instructions/vehicle fault		Mileage			

Work carried out and recommendations (include PPE & special precautions taken)

Parts and labour	Price
Total	

Data and specifications used (include the actual figures)

Assessor report

	Assessment outcome	Passed (tick ✓)
1	The learner worked safely and minimised risks to themselves and others	
2	The learner correctly selected and used appropriate technical information	
3	The learner correctly selected and used appropriate tools and equipment	
4	The learner correctly carried out the task required using suitable methods and testing procedures	
5	The learner correctly recorded information and made suitable recommendations	

	Tick	Written feedback (with reference to assessment criteria) must be given when a learner is referred
Pass: I confirm that the learner's work was to an acceptable standard and met the assessment criteria of the unit		
Refer: The work carried out did not achieve the standards specified by the assessment criteria		

Assessor name (print)	Assessor PIN/ref.	Date

Section below only to be completed by the learner once the assessor decision has been made and feedback given			
I confirm that the work carried out was my own, and that I received feedback from the Assessor	Learner name (print)	Learner signature	Date

Worksheet 10: Drain and top up coolant

Procedure

▶ Obtain information from the driver on the frequency of topping off the coolant. Check coolant level. Check the water pump drive belt for condition and tension.

▶ Pressure test the cooling system and the radiator cap. Look for leakage, condition of hoses, the pressure cap operating pressure, seals and vacuum valve. Report any defects and recommend rectification before adding new coolant.

▶ Identify the coolant drain points in the radiator and engine block. Undo and drain coolant into a clean drain tray. Dispose of old coolant in accordance with local environmental regulations. Close the radiator and engine block drain taps or plugs.

▶ Mix new coolant to the specified mixture of water and antifreeze with inhibitors or use a premixed coolant. Do not exceed a 50% to 60% ethylene glycol content as this does not improve the frost protection. (Over 70% reduces the protection).

▶ Add coolant to the radiator until the system is full. Open bleed screws to allow air to flow from the system.

▶ Run the engine and bleed air from the heater or high point in the system if required. Follow the engine manufacturer's instructions. Visually check for leaks.

▶ Road test and check the operation of the cooling system and the heater temperature range. Check the operation of the engine coolant temperature gauge.

▶ After road testing, visually recheck the system for leaks.

▶ Allow the engine to cool and check the coolant level. Top up if necessary. Do not overfill.

Job card

Technician/learner name & date	Make and model	VIN no.	Reg. no.	Job/task no.

Customer's instructions/vehicle fault	Mileage	

Work carried out and recommendations (include PPE & special precautions taken)

Parts and labour	Price
Total	

Data and specifications used (include the actual figures)

Assessor report

	Assessment outcome	Passed (tick ✓)
1	The learner worked safely and minimised risks to themselves and others	
2	The learner correctly selected and used appropriate technical information	
3	The learner correctly selected and used appropriate tools and equipment	
4	The learner correctly carried out the task required using suitable methods and testing procedures	
5	The learner correctly recorded information and made suitable recommendations	

	Tick	Written feedback (with reference to assessment criteria) must be given when a learner is referred
Pass: I confirm that the learner's work was to an acceptable standard and met the assessment criteria of the unit		
Refer: The work carried out did not achieve the standards specified by the assessment criteria		

Assessor name (print)	Assessor PIN/ref.	Date

Section below only to be completed by the learner once the assessor decision has been made and feedback given			
I confirm that the work carried out was my own, and that I received feedback from the Assessor	Learner name (print)	Learner signature	Date

Worksheet 11: Inspect water pump

Procedure

▶ Disconnect battery earth or ground lead. Drain the coolant into a clean drain tray. Remove the water pump. Look at the general condition, corrosion, cracks, sealing or other faults.

▶ Check the spindle bearings for wear by rotating and rocking the spindle. Feel and listen for roughness, grating and other abnormalities.

▶ Look closely at the water pump drive pulley flange to spindle joint, usually an interference fit, for signs of slippage.

▶ Look closely at the impeller to spindle joint, also an interference fit, for signs of slippage. Grip the drive flange and impeller and attempt to twist in opposite directions to feel for slippage.

▶ Clean the mating faces of the water pump and engine block and measure for flatness (true). Check bypass drillings are clean and clear.

▶ Report faults found.

▶ Reassemble water pump to engine.

Job card

Technician/learner name & date	Make and model	VIN no.		Reg. no.	Job/task no.

Customer's instructions/vehicle fault	Mileage	

Work carried out and recommendations (include PPE & special precautions taken)

Parts and labour	Price
Total	

Data and specifications used (include the actual figures)

Assessor report

Assessment outcome		Passed (tick ✓)
1	The learner worked safely and minimised risks to themselves and others	
2	The learner correctly selected and used appropriate technical information	
3	The learner correctly selected and used appropriate tools and equipment	
4	The learner correctly carried out the task required using suitable methods and testing procedures	
5	The learner correctly recorded information and made suitable recommendations	

	Tick	Written feedback (with reference to assessment criteria) must be given when a learner is referred
Pass: I confirm that the learner's work was to an acceptable standard and met the assessment criteria of the unit		
Refer: The work carried out did not achieve the standards specified by the assessment criteria		

Assessor name (print)	Assessor PIN/ref.	Date

Section below only to be completed by the learner once the assessor decision has been made and feedback given			
I confirm that the work carried out was my own, and that I received feedback from the Assessor	Learner name (print)	Learner signature	Date

Worksheet 12: Inspect drive belt condition and tension, inspect water pump bearings and seal for wear and leakage, check operation of cooling fan and airflow through radiator

Procedure

▶ Disconnect the battery earth or ground lead. Check the coolant level (cold engine). Identify the drive belt for the water pump – inside the engine compartment or under the vehicle. Check the condition of the belt, look for frays, cracks, glazing of the drive faces caused by slippage and check that the tension or free play is within manufacturer's tolerances. Free play of about 12 mm (½ inch) on the long side of a V belt is usually correct.

▶ Rock the water pump at the pulley and feel for wear/free play in the spindle bearings. Look for signs of leakage at the water pump spindle seal and the housing to engine block gasket. Look under the pump for tell-tale stains of coolant leakage on and under the pump. For engine-driven and electrically driven cooling fans check for damage and security. Check the fan cowling for damage and security. This is particularly important for electrically driven fans where the cowling is also the support bracket for the motor and fan assembly.

▶ Check the radiator matrix air passages for blockage. Blow through with an air line if a build-up of dust or dirt is visible. Blow through in the reverse direction. For engine-driven fans with a viscous thermostatic coupling check that the fan is free on the hub when the engine is cold. For all fans check that the fitting is correct. Some old steel fans could be fitted the wrong way round. For electrically driven fans check the motor spindle bearings by rocking and rotating the fan blades and feeling for free play, lumpiness and grinding in the bearings. (Ensure that the battery earth or ground lead is disconnected whenever handling electrically driven fans.)

▶ Reconnect the battery and run the engine. Check for airflow from the fan and a good air draw through the radiator. For electrically driven fans run the engine to the switch temperature so that the fan operates.

▶ Do not allow engine to overheat if the fan switch does not operate or the fan fails to operate. Check the fan motor by disconnecting the switch terminal block and bridging the connections. The motor should run.

▶ If the motor does not run check for an electrical feed at the terminal block. Connect a test lamp or voltmeter between each terminal and a good earth or ground – turn on the ignition if necessary. Check the switch continuity when cold (not continuous) and when hot (continuous).

Job card

Technician/learner name & date	Make and model	VIN no.		Reg. no.	Job/task no.

Customer's instructions/vehicle fault		Mileage	

Work carried out and recommendations (include PPE & special precautions taken)

Parts and labour	Price
Total	

Data and specifications used (include the actual figures)

Assessor report

	Assessment outcome	Passed (tick ✓)
1	The learner worked safely and minimised risks to themselves and others	
2	The learner correctly selected and used appropriate technical information	
3	The learner correctly selected and used appropriate tools and equipment	
4	The learner correctly carried out the task required using suitable methods and testing procedures	
5	The learner correctly recorded information and made suitable recommendations	

	Tick	Written feedback (with reference to assessment criteria) must be given when a learner is referred
Pass: I confirm that the learner's work was to an acceptable standard and met the assessment criteria of the unit		
Refer: The work carried out did not achieve the standards specified by the assessment criteria		

Assessor name (print)	Assessor PIN/ref.	Date

Section below only to be completed by the learner once the assessor decision has been made and feedback given			
I confirm that the work carried out was my own, and that I received feedback from the Assessor	Learner name (print)	Learner signature	Date

Worksheet 13: Remove and replace thermostat

Procedure

▶ Disconnect battery earth or ground lead. Drain the coolant into a clean drain tray.

▶ Undo the bolts holding the thermostat housing and remove the bolts and housing. For thermostats fitted in hoses or integrated in housings remove the hose or housing.

▶ Lift out the thermostat and inspect the thermostat housing gasket and the thermostat seal for signs of leakage and other deterioration. Test the thermostat if required.

▶ Clean and inspect the thermostat housing and cover mating faces. Inspect for flatness with a straight edge or surface table. Rectify if necessary.

▶ Fit the thermostat into the housing with new seals, if required. Fit the cover with a new gasket and sealant, if recommended, or grease. Fit the securing bolts with a thread sealant if they pass into the water passages and tighten to the specified torque.

▶ Refill the cooling system. Top up with a correct water and antifreeze mixture to make good any spilt or lost coolant during draining. Run the engine and bleed air from the system if necessary.

▶ Road test and check the engine temperature gauge and the heater operation for hot, cold and intermediate settings. Check airflow temperature agrees with heater settings.

▶ After road testing, visually recheck the system for leaks.

▶ Allow the engine to cool and check the coolant level. Top up if necessary. Do not overfill.

Job card

Technician/learner name & date	Make and model	VIN no.	Reg. no.	Job/task no.

Customer's instructions/vehicle fault		Mileage	

Work carried out and recommendations (include PPE & special precautions taken)

Parts and labour	Price
Total	

Data and specifications used (include the actual figures)

Assessor report

	Assessment outcome	Passed (tick ✓)
1	The learner worked safely and minimised risks to themselves and others	
2	The learner correctly selected and used appropriate technical information	
3	The learner correctly selected and used appropriate tools and equipment	
4	The learner correctly carried out the task required using suitable methods and testing procedures	
5	The learner correctly recorded information and made suitable recommendations	

	Tick	Written feedback (with reference to assessment criteria) must be given when a learner is referred
Pass: I confirm that the learner's work was to an acceptable standard and met the assessment criteria of the unit		
Refer: The work carried out did not achieve the standards specified by the assessment criteria		

Assessor name (print)	Assessor PIN/ref.	Date

Section below only to be completed by the learner once the assessor decision has been made and feedback given			
I confirm that the work carried out was my own, and that I received feedback from the Assessor	Learner name (print)	Learner signature	Date

Worksheet 14: Inspect cooling system, pressure test, check coolant condition and antifreeze

Procedure

▶ Obtain a suitable cooling system pressure tester and an ethylene glycol coolant hydrometer. Remove the radiator or expansion tank cap. Check the coolant for level, colour and contamination such as oil or rust and dirt.

▶ Check the ethylene glycol content with the hydrometer – adjust for temperature – and compare with manufacturer's specifications. Usually between 25% and 50% depending on the area of vehicle operation.

▶ Top up the coolant if necessary and attach the pressure tester using a suitable adapter in the filler neck.

▶ Apply a pressure equal to the system operating pressure, which is shown on the radiator filler cap – but should be checked from the manufacturer's data. *Do not exceed the stated pressure.* Observe the reading on the tester pressure gauge – it should remain stable. If the gauge reading drops, look for a good connection to the system and then for leaks in the system.

▶ Look at all hoses, pipes, joints, gaskets, the water pump, the heater and water valve for external leaks. Look inside the vehicle under the heater for leaks from the heater matrix. If no external leaks are visible, check the coolant for oil contamination. (Possible internal leaks.) Apply a pressure of about half the operating pressure and run the engine. Look for a rapid pressure rise, which would indicate a cylinder head gasket leak or crack between the cylinder or cylinder head and the water jacket.

▶ If the pressure remains stable but internal leakage is suspected, use a combustion gas detector kit with the pressure tester. If the test fluid changes colour during the test combustion, gases have been detected in the coolant confirming an internal leak.

▶ Fit the radiator/expansion tank pressure cap to the tester using an appropriate adapter and check that the operating pressure is held at the specified value. Remove the cap and check the vacuum valve seal operation and condition. Compare all findings with manufacturer's data. Report faults found.

Job card

Technician/learner name & date	Make and model	VIN no.		Reg. no.	Job/task no.

Customer's instructions/vehicle fault		Mileage	

Work carried out and recommendations (include PPE & special precautions taken)

Parts and labour	Price
Total	

Data and specifications used (include the actual figures)

Assessor report

	Assessment outcome	Passed (tick ✓)
1	The learner worked safely and minimised risks to themselves and others	
2	The learner correctly selected and used appropriate technical information	
3	The learner correctly selected and used appropriate tools and equipment	
4	The learner correctly carried out the task required using suitable methods and testing procedures	
5	The learner correctly recorded information and made suitable recommendations	

	Tick	Written feedback (with reference to assessment criteria) must be given when a learner is referred
Pass: I confirm that the learner's work was to an acceptable standard and met the assessment criteria of the unit		
Refer: The work carried out did not achieve the standards specified by the assessment criteria		

Assessor name (print)	Assessor PIN/ref.	Date

Section below only to be completed by the learner once the assessor decision has been made and feedback given			
I confirm that the work carried out was my own, and that I received feedback from the Assessor	Learner name (print)	Learner signature	Date

Worksheet 15: Remove and replace radiator and electric fan motor and switches

Procedure

▶ Disconnect battery earth or ground lead. Drain radiator coolant into a clean drain tray. Undo and remove top, bottom and expansion tank hose clips and pull off hoses. Disconnect the electrical terminal block to the motor and switch. Leave the cowl, motor and switch in place. Remove the radiator and cowl as an assembly.

▶ For radiators with integral transmission oil cooling, clamp the feed and return hoses, undo the hose clips and pull off the hoses. Catch the lost oil in a clean drain tray. OR undo the union nuts on steel pipes and remove the pipes. Catch the lost oil in a drain tray.

▶ For radiators with an air conditioning radiator or condenser attached to the cooling radiator, undo the attaching screws and support the condenser. *Do not undo the air conditioner refrigerant pipes unless the system has been discharged* (specialist operation).

▶ Undo and remove the radiator securing bolts and remove with the brackets. Carefully lift out the radiator assembly. Detach the cowl, motor and fan from the radiator. Mark all bolts or screws for replacement in exactly the same places. Unscrew the switch from the radiator.

▶ Reassemble in the reverse order. Ensure that the correct bolts or screws are replaced in their original positions. Some bolts may be longer than others and can puncture the radiator if fitted incorrectly.

▶ Refill the cooling system. Top up with a correct water and antifreeze mixture to make good any spilt or lost coolant during draining. Run the engine and bleed air from the system if necessary.

▶ Road test and check the engine temperature gauge, the heater operation for hot, cold and intermediate settings. Check airflow temperature agrees with heater settings.

▶ After road testing, visually recheck the system for leaks.

▶ Allow the engine to cool and check the coolant level. Top up if necessary. Do not overfill.

Job card

Technician/learner name & date	Make and model	VIN no.		Reg. no.	Job/task no.

Customer's instructions/vehicle fault		Mileage	

Work carried out and recommendations (include PPE & special precautions taken)

Parts and labour	Price
Total	

Data and specifications used (include the actual figures)

Assessor report

	Assessment outcome	Passed (tick ✓)
1	The learner worked safely and minimised risks to themselves and others	
2	The learner correctly selected and used appropriate technical information	
3	The learner correctly selected and used appropriate tools and equipment	
4	The learner correctly carried out the task required using suitable methods and testing procedures	
5	The learner correctly recorded information and made suitable recommendations	

	Tick	Written feedback (with reference to assessment criteria) must be given when a learner is referred
Pass: I confirm that the learner's work was to an acceptable standard and met the assessment criteria of the unit		
Refer: The work carried out did not achieve the standards specified by the assessment criteria		

Assessor name (print)	Assessor PIN/ref.	Date

Section below only to be completed by the learner once the assessor decision has been made and feedback given			
I confirm that the work carried out was my own, and that I received feedback from the Assessor	Learner name (print)	Learner signature	Date

Worksheet 16: Reverse flush and test radiator flow rate. Inspect hoses

Procedure

▶ Disconnect battery earth or ground lead. Inspect hoses for security and condition. Inspect the hoses systematically starting at the top hose. Look at the hose for signs of leakage, tell-tale markings at each end, and for cracking, perishing, chaffing and other damage or deterioration. Squeeze the hose to feel for internal damage from perishing. If internal damage is suspected remove the hose for internal inspection. Look for a break-up of the inner lining of the hose and particles of old hose in the coolant. Check all other hoses, bottom hose, heater hoses, bypass hose and any other coolant hoses manifolds or pipes.

▶ If the radiator and cooling system is to be cleaned by reverse flushing, drain the coolant into a clean drain tray. Undo the bottom hose clip at the water pump inlet and pull off the hose. Inspect the hose internally for condition. Replace the hose if defective.

▶ Remove the radiator pressure cap. Push a mains water hose into the bottom hose and plug with rag to seal the end. Turn on the mains water and allow the water to flow from the radiator header tank filler neck until it is clean and flows freely. A poor flow indicates a blocked radiator.

▶ In order to clean the engine block by reverse flushing, remove the thermostat and refit the housing cover. Refit the radiator cap and apply mains water pressure to the bottom hose. Allow the water to flow until it runs clean and freely from the water pump bottom hose connection.

▶ Refit the thermostat with a new gasket and sealant on the gasket and any bolts that run into the water passages. Refit all hoses and tighten the clips. Mix new coolant to the specified mixture of water and antifreeze with inhibitors or use a premixed coolant. Do not exceed a 50% to 60% ethylene glycol content, as this does not improve the frost protection. (Over 70% reduces the protection.)

▶ Refill the cooling system. Run the engine and bleed air from the system if necessary. Road test and check the engine temperature gauge, and the heater operation for hot, cold and intermediate settings. Check airflow temperature agrees with heater settings.

▶ After road testing visually recheck the system for leaks Allow the engine to cool and check the coolant level. Top up if necessary. Do not overfill.

Job card

Technician/learner name & date	Make and model	VIN no.		Reg. no.	Job/task no.

Customer's instructions/vehicle fault		Mileage	

Work carried out and recommendations (include PPE & special precautions taken)

Parts and labour	Price
Total	

Data and specifications used (include the actual figures)

Assessor report

	Assessment outcome	Passed (tick ✓)
1	The learner worked safely and minimised risks to themselves and others	
2	The learner correctly selected and used appropriate technical information	
3	The learner correctly selected and used appropriate tools and equipment	
4	The learner correctly carried out the task required using suitable methods and testing procedures	
5	The learner correctly recorded information and made suitable recommendations	

	Tick	Written feedback (with reference to assessment criteria) must be given when a learner is referred
Pass: I confirm that the learner's work was to an acceptable standard and met the assessment criteria of the unit		
Refer: The work carried out did not achieve the standards specified by the assessment criteria		

Assessor name (print)	Assessor PIN/ref.	Date

Section below only to be completed by the learner once the assessor decision has been made and feedback given			
I confirm that the work carried out was my own, and that I received feedback from the Assessor	Learner name (print)	Learner signature	Date

Worksheet 17: Routine maintenance inspections and replacement of cooling parts

Procedure

▶ Visual/coolant level (safety – do not remove cap from hot engine).

▶ Measure antifreeze content with hydrometer (safety – avoid skin contact with coolant).

▶ Visual/all hoses for leakage and condition.

▶ Visual/all gaskets for leakage.

▶ Check with pressure tester if leaks suspected (do not exceed system rated pressure).

▶ Check pressure cap operating pressure. Inspect vacuum valve.

▶ Check water pump drive belt condition and tension. Adjust if necessary.

▶ Check operation of cooling fan and airflow through the radiator. Clean if necessary.

▶ Check water pump bearing for free play. Listen for abnormal noises.

▶ Check operation of heater, heater controls and blower motor.

▶ At regular intervals replace coolant with new water and antifreeze solution.

▶ Additional items – see manufacturer's schedule.

Job card

Technician/learner name & date	Make and model	VIN no.		Reg. no.	Job/task no.

Customer's instructions/vehicle fault		Mileage			

Work carried out and recommendations (include PPE & special precautions taken)

Parts and labour	Price
Total	

Data and specifications used (include the actual figures)

Assessor report

	Assessment outcome	Passed (tick ✓)
1	The learner worked safely and minimised risks to themselves and others	
2	The learner correctly selected and used appropriate technical information	
3	The learner correctly selected and used appropriate tools and equipment	
4	The learner correctly carried out the task required using suitable methods and testing procedures	
5	The learner correctly recorded information and made suitable recommendations	

	Tick	Written feedback (with reference to assessment criteria) must be given when a learner is referred
Pass: I confirm that the learner's work was to an acceptable standard and met the assessment criteria of the unit		
Refer: The work carried out did not achieve the standards specified by the assessment criteria		

Assessor name (print)	Assessor PIN/ref.	Date

Section below only to be completed by the learner once the assessor decision has been made and feedback given			
I confirm that the work carried out was my own, and that I received feedback from the Assessor	Learner name (print)	Learner signature	Date

Worksheet 18: Remove and replace hoses including radiator, heater and bypass hoses

Procedure

▶ Disconnect battery ground lead. Drain the coolant into a clean drain tray.

▶ Undo the hose clips and slide along the hose. If the hose is not free on its connector insert a blunt blade (screwdriver or special tool) under the lip of the hose and ease round to break the seal.

▶ Pull the hose off from the connector. Take care with radiator couplings as these can be damaged by the use of excessive force.

▶ Transfer or fit new hose clips to the hose for replacement. Fit the new hoses with a sealant on the joint if recommended, otherwise a dry joint will give a good seal. Slide the hose clips into place and tighten.

▶ Refill the cooling system. Top up with a correct water and antifreeze mixture to make good any spilt or lost coolant during draining. Run the engine and bleed air from the system if necessary.

▶ Road test and check the engine temperature gauge, the heater operation for hot, cold and intermediate settings. Check airflow temperature agrees with heater settings.

▶ After road testing, visually recheck the system for leaks.

▶ Allow the engine to cool and check the coolant level. Top up if necessary. Do not overfill.

Job card

Technician/learner name & date	Make and model	VIN no.		Reg. no.	Job/task no.

Customer's instructions/vehicle fault		Mileage	

Work carried out and recommendations (include PPE & special precautions taken)

Parts and labour	Price
Total	

Data and specifications used (include the actual figures)

Assessor report

	Assessment outcome	Passed (tick ✓)
1	The learner worked safely and minimised risks to themselves and others	
2	The learner correctly selected and used appropriate technical information	
3	The learner correctly selected and used appropriate tools and equipment	
4	The learner correctly carried out the task required using suitable methods and testing procedures	
5	The learner correctly recorded information and made suitable recommendations	

	Tick	Written feedback (with reference to assessment criteria) must be given when a learner is referred
Pass: I confirm that the learner's work was to an acceptable standard and met the assessment criteria of the unit		
Refer: The work carried out did not achieve the standards specified by the assessment criteria		

Assessor name (print)	Assessor PIN/ref.	Date

Section below only to be completed by the learner once the assessor decision has been made and feedback given			
I confirm that the work carried out was my own, and that I received feedback from the Assessor	Learner name (print)	Learner signature	Date

Worksheet 19: Drain and top up coolant for winter usage or replace coolant

Procedure

▶ Obtain information from the driver on the frequency of topping up the coolant. Check coolant level. Check the water pump drive belt for condition and tension.

▶ Pressure test the cooling system and the radiator cap. Look for leakage, condition of hoses, the pressure cap operating pressure, seals and vacuum valve. Report any defects and recommend rectification before adding new coolant.

▶ Identify the coolant drain points in the radiator and engine block. Undo and drain coolant into a clean drain tray. Dispose of old coolant in accordance with local environmental regulations. Close the radiator and engine block drain taps or plugs.

▶ Mix new coolant to the specified mixture of water and antifreeze with inhibitors or use a premixed coolant. Do not exceed a 50% to 60% ethylene glycol content as this does not improve the frost protection. (Over 70% reduces the protection).

▶ Add coolant to the radiator until the system is full. Open bleed screws to allow air to flow from the system.

▶ Run the engine and bleed air from the heater or high point in the system if required. Follow the engine manufacturer's instructions. Visually check for leaks.

▶ Road test and check the operation of the cooling system and the heater temperature range. Check the operation of the engine coolant temperature gauge.

▶ After road testing, visually recheck the system for leaks.

▶ Allow the engine to cool and check the coolant level. Top up if necessary. Do not overfill.

Job card

Technician/learner name & date	Make and model	VIN no.		Reg. no.	Job/task no.

Customer's instructions/vehicle fault		Mileage			

Work carried out and recommendations (include PPE & special precautions taken)

Parts and labour	Price
Total	

Data and specifications used (include the actual figures)

Assessor report

	Assessment outcome	Passed (tick ✓)
1	The learner worked safely and minimised risks to themselves and others	
2	The learner correctly selected and used appropriate technical information	
3	The learner correctly selected and used appropriate tools and equipment	
4	The learner correctly carried out the task required using suitable methods and testing procedures	
5	The learner correctly recorded information and made suitable recommendations	

	Tick	Written feedback (with reference to assessment criteria) must be given when a learner is referred
Pass: I confirm that the learner's work was to an acceptable standard and met the assessment criteria of the unit		
Refer: The work carried out did not achieve the standards specified by the assessment criteria		

Assessor name (print)	Assessor PIN/ref.	Date

Section below only to be completed by the learner once the assessor decision has been made and feedback given			
I confirm that the work carried out was my own, and that I received feedback from the Assessor	Learner name (print)	Learner signature	Date

Worksheet 20: Inspect, clean or renew and gap spark plugs

Procedure

▶ Check age/mileage of spark plugs from customer records. Check manufacturer's data for correct plugs for vehicle.

▶ Check security of spark plug leads during disconnection – note and rectify any loose connections. Using an appropriate deep socket to remove the spark plugs. Note any tight threads and rectify before reassembly. Lay out plugs in order for diagnosis of plug or engine condition.

▶ Inspect plugs for condition – erosion of electrodes, carbon fouling, damaged insulation, thread condition and tightness of threaded terminals. Check sealing washer or taper seat condition.

▶ Look also for symptoms of engine or fuel system faults. Black sooty deposits indicate a rich fuel mixture, black oily deposits indicate piston ring or inlet valve stem wear, white/brown sooty deposits indicate a weak mixture and overheating. Investigate faults found, replace plugs.

▶ Plugs in good condition can be cleaned and re-gapped with a feeler gauge. All new or replaced plugs should be gapped to the manufacturer's specification before fitting.

▶ Adjust the earth/ground electrode to give a gap between the centre and earth/ground electrodes. *Do not* exert a force on the centre electrode, either lever out with a suitable tool to increase the gap or tap in on a solid object to reduce the gap. Check with a feeler or gap gauge.

▶ Lubricate the threads of plugs being fitted into aluminium cylinder heads (graphite or high melting point grease as specified by the manufacturer). Fit into cylinder head and tighten hand tight before finally tightening with a torque wrench. Ensure correct torque for washer or taper seat types.

▶ Reconnect leads and check security – leads must be tight and making a good electrical connection.

▶ Run engine/road test and check for correct operation.

Job card

Technician/learner name & date	Make and model	VIN no.	Reg. no.	Job/task no.

Customer's instructions/vehicle fault		Mileage	

Work carried out and recommendations (include PPE & special precautions taken)

Parts and labour	Price
Total	

Data and specifications used (include the actual figures)

Assessor report

Assessment outcome	Passed (tick ✓)
1	The learner worked safely and minimised risks to themselves and others	
2	The learner correctly selected and used appropriate technical information	
3	The learner correctly selected and used appropriate tools and equipment	
4	The learner correctly carried out the task required using suitable methods and testing procedures	
5	The learner correctly recorded information and made suitable recommendations	

	Tick	Written feedback (with reference to assessment criteria) must be given when a learner is referred
Pass: I confirm that the learner's work was to an acceptable standard and met the assessment criteria of the unit		
Refer: The work carried out did not achieve the standards specified by the assessment criteria		

Assessor name (print)	Assessor PIN/ref.	Date

Section below only to be completed by the learner once the assessor decision has been made and feedback given			
I confirm that the work carried out was my own, and that I received feedback from the Assessor	Learner name (print)	Learner signature	Date

Worksheet 21: Inspect fuel system for leaks and condition of pipes and hoses, etc.

Procedure

▶ This is a visual inspection. Check for fuel odour under the vehicle and in the engine compartment.

▶ Raise the vehicle onto axle stands or a vehicle hoist. Open the bonnet. Follow the fuel lines from the tank to the carburettor of fuel injection components. Look carefully at all pipes and hoses. Look from the filler neck to the tank, at the tank, and the feed and return pipes.

▶ Look for washed areas, stains and fuel odour and any other signs of leakage. Check for damage, routing and chaffing. Look for corrosion, perishing, leaks at joints and security of hose clips and pipe securing clips.

▶ Look at the fuel tank security, and for leaks from the fuel gauge sender unit gasket and outlet pipe.

▶ Inspect the vapour lines, vapour trap and the filler cap fit and condition of the sealing ring. Modern tanks hold a small pressure or vacuum under some conditions. There is often a rush of air as the filler cap is removed – this is normal.

▶ Report any defects found.

Job card

Technician/learner name & date	Make and model	VIN no.	Reg. no.	Job/task no.

Customer's instructions/vehicle fault	Mileage	

Work carried out and recommendations (include PPE & special precautions taken)

Parts and labour	Price
Total	

Data and specifications used (include the actual figures)

Assessor report

	Assessment outcome	Passed (tick ✓)
1	The learner worked safely and minimised risks to themselves and others	
2	The learner correctly selected and used appropriate technical information	
3	The learner correctly selected and used appropriate tools and equipment	
4	The learner correctly carried out the task required using suitable methods and testing procedures	
5	The learner correctly recorded information and made suitable recommendations	

	Tick	Written feedback (with reference to assessment criteria) must be given when a learner is referred
Pass: I confirm that the learner's work was to an acceptable standard and met the assessment criteria of the unit		
Refer: The work carried out did not achieve the standards specified by the assessment criteria		

Assessor name (print)	Assessor PIN/ref.	Date

Section below only to be completed by the learner once the assessor decision has been made and feedback given			
I confirm that the work carried out was my own, and that I received feedback from the Assessor	Learner name (print)	Learner signature	Date

Worksheet 22: Check exhaust system for condition, leaks, blockage and security

Procedure

▶ Open bonnet – run engine and then accelerate and listen for abnormal noises – exhaust blow, squeal, screech, excessive noise level – and that the engine revs freely (possible blockage).

▶ Allow the engine to idle – cover the tail pipe outlet with a cloth pad – listen for exhaust blow.

▶ Inspect under the bonnet for exhaust manifold gasket leaks – look for tell-tale black markings from a loose or broken joint.

▶ Under the vehicle follow the exhaust pipes along – look at the down/front pipe to exhaust manifold joint and all the pipe to pipe or pipe to silencer/muffler/catalyst/resonator/etc. joints. Look for tell-tale black markings from loose, broken or badly sealed joints.

▶ Look closely at the skin of all pipes, silencers, mufflers, catalysts, resonators, etc. Look for corrosion, holes or other damage or deterioration.

▶ Look closely at the fitting, security, condition and positioning of heat shields. These must be in place to protect flammable materials inside the vehicle (sound deadening, etc.) from ignition from the high temperature of exhausts and catalytic converters.

▶ Look closely at all exhaust mounting brackets and rubber mounting components (hangers) for fitting, corrosion, perishing, separation, tension, etc. On turbocharged engines check the down/front pipe support bracket for condition and security.

▶ Check the position and routing of the exhaust for knocking on the body or chassis and for fouling on other components such as brake pipes, steering, suspension, axle, etc.

▶ If blockage is suspected, exhaust gas leaks in front of the blockage are likely. Disconnect a pipe at a suitable position in front of the blockage and check the flow through the remainder of the system with an air line blower – plug the pipe around the blower with a cloth pad.

Job card

Technician/learner name & date	Make and model	VIN no.		Reg. no.	Job/task no.

Customer's instructions/vehicle fault	Mileage	

Work carried out and recommendations (include PPE & special precautions taken)

Parts and labour	Price
Total	

Data and specifications used (include the actual figures)

Assessor report

Assessment outcome		Passed (tick ✓)
1	The learner worked safely and minimised risks to themselves and others	
2	The learner correctly selected and used appropriate technical information	
3	The learner correctly selected and used appropriate tools and equipment	
4	The learner correctly carried out the task required using suitable methods and testing procedures	
5	The learner correctly recorded information and made suitable recommendations	

	Tick	Written feedback (with reference to assessment criteria) must be given when a learner is referred
Pass: I confirm that the learner's work was to an acceptable standard and met the assessment criteria of the unit		
Refer: The work carried out did not achieve the standards specified by the assessment criteria		

Assessor name (print)		Assessor PIN/ref.	Date

Section below only to be completed by the learner once the assessor decision has been made and feedback given			
I confirm that the work carried out was my own, and that I received feedback from the Assessor	Learner name (print)	Learner signature	Date

Worksheet 23: Routine maintenance inspections, lubrication and replacement of parts

Procedure

▶ General visual inspection – listen for abnormal noises.

▶ Check operation of ignition/generator warning light.

▶ Listen to operation of starter motor.

▶ Check battery electrolyte level – top up if necessary with specially formulated water.

▶ Check battery casing for leaks and battery security.

▶ Check alternator drive belt for condition and tension.

▶ Check all electrical cables to battery, starter and alternator.

▶ Coat battery terminals with petroleum jelly.

Job card

Technician/learner name & date	Make and model	VIN no.		Reg. no.	Job/task no.

Customer's instructions/vehicle fault	Mileage	

Work carried out and recommendations (include PPE & special precautions taken)

	Price
Parts and labour	
Total	

Data and specifications used (include the actual figures)

Assessor report

	Assessment outcome	*Passed (tick ✓)*
1	The learner worked safely and minimised risks to themselves and others	
2	The learner correctly selected and used appropriate technical information	
3	The learner correctly selected and used appropriate tools and equipment	
4	The learner correctly carried out the task required using suitable methods and testing procedures	
5	The learner correctly recorded information and made suitable recommendations	

	Tick	Written feedback (with reference to assessment criteria) must be given when a learner is referred
Pass: I confirm that the learner's work was to an acceptable standard and met the assessment criteria of the unit		
Refer: The work carried out did not achieve the standards specified by the assessment criteria		

Assessor name (print)	Assessor PIN/ref.	Date

Section below only to be completed by the learner once the assessor decision has been made and feedback given			
I confirm that the work carried out was my own, and that I received feedback from the Assessor	Learner name (print)	Learner signature	Date

Worksheet 24: Check battery and charge circuit operation for winter usage of vehicle

Procedure

▶ Check the alternator drive belt for condition and tension. Look for frays, cracks, glazing of the drive faces caused by slippage and check that the tension or free play is about 12 mm (½ in).

▶ Check the security of the battery and the condition of the casing. Look for signs of corrosion or other deterioration on the straps/clamps, on and around the battery and on the battery cables at the battery and at where they connect to the body and to the starter system.

▶ Disconnect both battery leads, (earth or ground first), and clean the battery terminals and the cable terminals.

▶ Check the state of charge of the battery with a hydrometer and/or digital voltmeter. Hydrometer readings should be near fully charged (1.28) and the same for all cells. *Or*, check the built-in hydrometer of maintenance-free batteries, which should be green for a fully charged battery and black if partially charged. A yellow dot indicates that the battery is defective. Voltage readings should be 13 volts for conventional lead acid batteries and 12.8 volts for maintenance-free batteries.

▶ If hydrometer and voltage readings are low, recharge the battery. If the readings are correct, carry out a high rate discharge test. Connect the tester and set the discharge rate to the manufacturer's recommended setting. *Do not* exceed 3x the Ah rate (10 hour rate) and 15 second duration.

▶ Reconnect the battery and coat the terminals with petroleum jelly. Check charge circuit cables for condition and that all terminals are clean and tight. Treat with petroleum jelly or water-repellent spray.

▶ Connect a digital voltmeter to the battery terminals and a clamp on ammeter to the alternator output lead.

▶ Start and run the engine at about 2000 rpm and check the voltage and amps readings. Turn on high consumption devices, head lamps, heater blower, heated windscreens, etc. and read volts and amps. Compare readings with manufacturer's data (volts 14.7, amps 20 +).

Job card

Technician/learner name & date	Make and model	VIN no.		Reg. no.	Job/task no.

Customer's instructions/vehicle fault	Mileage	

Work carried out and recommendations (include PPE & special precautions taken)

Parts and labour	Price
Total	

Data and specifications used (include the actual figures)

Assessor report

	Assessment outcome	Passed (tick ✓)
1	The learner worked safely and minimised risks to themselves and others	
2	The learner correctly selected and used appropriate technical information	
3	The learner correctly selected and used appropriate tools and equipment	
4	The learner correctly carried out the task required using suitable methods and testing procedures	
5	The learner correctly recorded information and made suitable recommendations	

	Tick	Written feedback (with reference to assessment criteria) must be given when a learner is referred
Pass: I confirm that the learner's work was to an acceptable standard and met the assessment criteria of the unit		
Refer: The work carried out did not achieve the standards specified by the assessment criteria		

Assessor name (print)	Assessor PIN/ref.	Date

Section below only to be completed by the learner once the assessor decision has been made and feedback given			
I confirm that the work carried out was my own, and that I received feedback from the Assessor	Learner name (print)	Learner signature	Date

Worksheet 25: Remove and replace battery, battery cables and securing devices

Procedure

▶ Open bonnet to access the battery. Check the general condition on and around the battery. Fit memory keeper (if available). Disconnect the earth or ground lead and then the supply lead from the battery.

▶ Undo and remove the battery-retaining straps, clamps and brackets and remove. Remove any obstructions that may prevent the battery being lifted out. Plan the removal so that the lift, route and new position of the battery are known beforehand. Keeping the battery as level as possible, lift up from the carrier and lift out from the vehicle. Place the battery on a bench or other suitable place.

▶ If it is necessary to use a lifting tool on the battery, ensure that it is correctly located on the battery casing, is secure and will not crush or otherwise damage the battery.

▶ Be very careful not to spill battery acid. If acid is split or has been lost from the battery, treat the area with a solution of baking soda and water or ammonia and water (an alkaline to neutralize the acid). Treat bare metal and repaint if necessary. Remove contaminated clothing and rinse as quickly as possible. Rinse acid from skin immediately. If acid burns are experienced, seek medical attention.

▶ Remove all cables by undoing securing bolts/nuts or pulling apart at the terminal or terminal block. Clean all terminals and connections to ensure good current flow.

▶ Reconnect and check the tightness of all push fit terminals. Coat with petroleum jelly or water repellent.

▶ Refit the battery, and if fitting a replacement, ensure that the new battery matches the old battery for casing dimensions, Ah capacity, and the type and position of the terminal posts. Fit retaining straps/clamps and tighten. Do not over tighten as battery damage can occur.

▶ Refit the battery supply lead and the earth or ground lead. Coat the terminals with petroleum jelly.

▶ Start and run the engine and check that the engine starter motor operated correctly and that the generator/ignition warning light came on and went out as the engine speed increased. Restore electronic memory functions if memory keeper not available.

Job card

Technician/learner name & date	Make and model	VIN no.		Reg. no.	Job/task no.

Customer's instructions/vehicle fault		Mileage			

Work carried out and recommendations (include PPE & special precautions taken)

Parts and labour	Price
Total	

Data and specifications used (include the actual figures)

Assessor report

	Assessment outcome	Passed (tick ✓)
1	The learner worked safely and minimised risks to themselves and others	
2	The learner correctly selected and used appropriate technical information	
3	The learner correctly selected and used appropriate tools and equipment	
4	The learner correctly carried out the task required using suitable methods and testing procedures	
5	The learner correctly recorded information and made suitable recommendations	

	Tick	Written feedback (with reference to assessment criteria) must be given when a learner is referred
Pass: I confirm that the learner's work was to an acceptable standard and met the assessment criteria of the unit		
Refer: The work carried out did not achieve the standards specified by the assessment criteria		

Assessor name (print)	Assessor PIN/ref.	Date

Section below only to be completed by the learner once the assessor decision has been made and feedback given			
I confirm that the work carried out was my own, and that I received feedback from the Assessor	Learner name (print)	Learner signature	Date

Worksheet 26: Inspect batteries for condition, security and state of charge

Procedure

▶ Open bonnet to access the battery. Check the general condition on and around the battery. Check the alternator drive belt for condition and tension. Look at the battery for signs of leakage from the casing. A build-up of a light coloured corrosive substance indicates the presence of battery acid. Look at the securing straps and brackets for security and condition. Look at the battery cables for condition, security and corrosion. Look at the earth or ground lead connection to the vehicle body for condition, security and corrosion. Make a general check of all cables to the starter and alternator and the engine earth or ground lead.

▶ Check the electrolyte level and specific gravity (relative density) with a hydrometer. *Or* check the built-in hydrometer on maintenance-free batteries. Green dot for charged, black for partially charged and yellow for internal faults where the battery should not be recharged. Carry out battery capacity test and confirm it has a capacity recommended by the vehicle manufacturer. Check the voltage with a digital voltmeter. Compare with the manufacturer's specifications. Conventional lead acid 12 volt batteries should read 13 to 13.2 volts when fully charged. Maintenance-free 12 volt batteries should read 12.8 volts when fully charged.

▶ Carry out a high rate discharge test *only* on a fully charged battery. Connect the tester following the manufacturer's instructions. Observe polarity. Set amperage to 3x battery Ah rate. Carry out test following the equipment manufacturer's instructions. *Do not* exceed test time. Conventional batteries can be 'fast charged' using a high output (amps) fast charger. Connect and charge at the current (amps) and time specified by the fast charger manufacturer. Disconnect the battery leads before connecting to the fast charger.

▶ *Do not* allow the battery to overheat. Use the temperature probe if fitted to the charger. *Do not* fast charge maintenance-free batteries unless permitted by the manufacturer. Slow (low current – amps) charging of conventional lead acid and maintenance-free batteries can be carried out singularly or two or more of a similar type and at a similar state of charge can be connected together.

▶ Connect a single battery to a charger observing the polarity of the charger and the battery. Red (+) to positive, black (-) to negative. Set the charger to 1/10th of the battery Ah (amp hour – 10 hour rating) and charge for 15 hours for a fully discharged battery. For multiple battery charging, connect batteries in parallel for a 12 volt charger. For 24 volt chargers, connect two 12 volt batteries in series. Four batteries are connected in series and parallel. Set the charge rate according to the number and Ah ratings of the batteries.

Job card

Technician/learner name & date	Make and model	VIN no.		Reg. no.	Job/task no.

Customer's instructions/vehicle fault		Mileage			

Work carried out and recommendations (include PPE & special precautions taken)

Parts and labour	Price
Total	

Data and specifications used (include the actual figures)

Assessor report

	Assessment outcome	Passed (tick ✓)
1	The learner worked safely and minimised risks to themselves and others	
2	The learner correctly selected and used appropriate technical information	
3	The learner correctly selected and used appropriate tools and equipment	
4	The learner correctly carried out the task required using suitable methods and testing procedures	
5	The learner correctly recorded information and made suitable recommendations	

	Tick	Written feedback (with reference to assessment criteria) must be given when a learner is referred
Pass: I confirm that the learner's work was to an acceptable standard and met the assessment criteria of the unit		
Refer: The work carried out did not achieve the standards specified by the assessment criteria		

Assessor name (print)	Assessor PIN/ref.	Date

Section below only to be completed by the learner once the assessor decision has been made and feedback given			
I confirm that the work carried out was my own, and that I received feedback from the Assessor	Learner name (print)	Learner signature	Date

Worksheet 27: Remove and replace drive belts and pulleys

Procedure

▶ Disconnect battery earth or ground lead. Identify the drive belt for the alternator – V, multi-vee (ribbed) or serpentine.

▶ Slacken the alternator securing bolts and drive belt tensioner bolts. Slacken the belt tension. Pull the belt from the alternator and other pulleys and remove. Inspect the belt and pulleys for signs of wear, damage and slipping.

▶ For automatic tensioners fit an appropriate socket and bar to the tensioner, pull anti-clockwise to release the tension. Hold the tensioner in the off position and pull off the belt. Slowly release the tensioner. *Do not* let it fly open under its own spring force as damage can occur.

▶ Fit the new belt and adjust the tension to the manufacturer's specification. For V belts the correct tension is usually about 13 mm or ½ inch of free play on the longest side. For multi-vee belts use a tension gauge or adjust so that the belt can be twisted through 90° on the long side.

▶ Tighten the tensioner securing bolts. Refit other components in reverse order. Run the engine and check for abnormal noises. A whine would indicate that the belt tension is too tight and a slapping noise of the belt against the cover indicates that the belt is too loose.

▶ To remove pulleys, undo the securing nuts and pull the pulley off its shaft. A special or two legged puller may be required. Remove the key from the shaft if necessary, and if replacing, it may be necessary to file to size to fit the pulley slot.

▶ To hold a pulley while undoing the securing nut, place an old drive belt around the pulley and hold the belt tightly in a vice whilst turning the nut with a wrench.

Job card

Technician/learner name & date	Make and model	VIN no.	Reg. no.	Job/task no.

Customer's instructions/vehicle fault		Mileage		

Work carried out and recommendations (include PPE & special precautions taken)

Parts and labour	Price
Total	

Data and specifications used (include the actual figures)

Assessor report

	Assessment outcome	Passed (tick ✓)
1	The learner worked safely and minimised risks to themselves and others	
2	The learner correctly selected and used appropriate technical information	
3	The learner correctly selected and used appropriate tools and equipment	
4	The learner correctly carried out the task required using suitable methods and testing procedures	
5	The learner correctly recorded information and made suitable recommendations	

	Tick	Written feedback (with reference to assessment criteria) must be given when a learner is referred
Pass: I confirm that the learner's work was to an acceptable standard and met the assessment criteria of the unit		
Refer: The work carried out did not achieve the standards specified by the assessment criteria		

Assessor name (print)	Assessor PIN/ref.	Date

Section below only to be completed by the learner once the assessor decision has been made and feedback given			
I confirm that the work carried out was my own, and that I received feedback from the Assessor	Learner name (print)	Learner signature	Date

Chassis

Worksheet 28: Service suspension system

Procedure

▶ The service requirements for suspension are very simple. Mostly, these involve quick checks for security and leaks. Dampers/shocks may, however, require replacement. Support the vehicle on a wheel-free ramp if possible.

Front damper/shock absorber

▶ Use an open-end wrench, or Allen key, to prevent the upper end from turning. Then remove the upper retaining nut. On some vehicles, it may be necessary to remove a wheel panel for better access. Remove the bolts retaining the lower shock absorber pivot to the control arm. Prior to installation, place the grommet and washer in position on the upper stem. Insert the stem through the upper mount, push the grommet and washer into place and install the retaining nut. Tighten the nut to the specified torque.

▶ Place the lower end of the shock absorber in position and install the bolts. Tighten the bolts to the specified torque.

Rear damper/shock absorber

▶ Use a backup wrench on the stud on the knuckle and remove the shock absorber nut, then remove the upper through bolt and nut and detach the shock absorber. To install, place the shock absorber in position and install the retaining nuts and bolts. Using a backup wrench on the stud, tighten the shock absorber-to-knuckle nut to the specified torque.

▶ Raise the knuckle to normal ride height with a jack and tighten the upper shock absorber through bolt and nut to the specified torque. Lubricate pivots and pins that are provided with grease nipples.

Job card

Technician/learner name & date	Make and model	VIN no.		Reg. no.	Job/task no.

Customer's instructions/vehicle fault		Mileage			

Work carried out and recommendations (include PPE & special precautions taken)

Parts and labour	Price
Total	

Data and specifications used (include the actual figures)

Assessor report

Assessment outcome		Passed (tick ✓)
1	The learner worked safely and minimised risks to themselves and others	
2	The learner correctly selected and used appropriate technical information	
3	The learner correctly selected and used appropriate tools and equipment	
4	The learner correctly carried out the task required using suitable methods and testing procedures	
5	The learner correctly recorded information and made suitable recommendations	

	Tick	Written feedback (with reference to assessment criteria) must be given when a learner is referred
Pass: I confirm that the learner's work was to an acceptable standard and met the assessment criteria of the unit		
Refer: The work carried out did not achieve the standards specified by the assessment criteria		

Assessor name (print)		Assessor PIN/ref.	Date

Section below only to be completed by the learner once the assessor decision has been made and feedback given			
I confirm that the work carried out was my own, and that I received feedback from the Assessor	Learner name (print)	Learner signature	Date

Worksheet 29: Check suspension and steering operation

Procedure

▶ Raise the front end of the vehicle and support it securely on stands placed under the frame rails. Because of the work to be done, make sure the vehicle cannot fall from the stands.

▶ Visually check the suspension and steering components for wear.

▶ Indications of a fault in these systems are excessive play in the steering wheel before the front wheels react, excessive sway around corners, body movement over rough roads or binding at some point as the steering wheel is turned.

▶ Check the wheel bearings. Do this by spinning the front wheels. Listen for any abnormal noises and watch to make sure the wheel spins true (does not wobble). Hold the top and bottom of the tyre; pull inward and then outward on the tyre. Note any movement, which would indicate a loose wheel bearing assembly. If the bearings are suspect, they should be replaced.

▶ From underneath the vehicle, check for loose bolts, broken or disconnected parts and deteriorated rubber bushings on all suspension and steering components. Look for grease or fluid leaking from the steering assembly. Check the power steering hoses and connections for leaks. Check the ball joints for wear by looking for movement as the steering is rocked.

▶ Have an assistant turn the steering wheel from side to side and check the steering components for free movement, chafing and binding. If the steering does not react with the movement of the steering wheel, try to determine where the freeplay is located.

▶ Steering systems use flexible rubber boots, which should be carefully checked for tears, oil contamination or damage.

▶ Lower the vehicle and report any faults found.

Job card

Technician/learner name & date	Make and model	VIN no.		Reg. no.	Job/task no.

Customer's instructions/vehicle fault		Mileage			

Work carried out and recommendations (include PPE & special precautions taken)

Parts and labour	Price
Total	

Data and specifications used (include the actual figures)

Assessor report

	Assessment outcome	Passed (tick ✓)
1	The learner worked safely and minimised risks to themselves and others	
2	The learner correctly selected and used appropriate technical information	
3	The learner correctly selected and used appropriate tools and equipment	
4	The learner correctly carried out the task required using suitable methods and testing procedures	
5	The learner correctly recorded information and made suitable recommendations	

	Tick	Written feedback (with reference to assessment criteria) must be given when a learner is referred
Pass: I confirm that the learner's work was to an acceptable standard and met the assessment criteria of the unit		
Refer: The work carried out did not achieve the standards specified by the assessment criteria		

Assessor name (print)	Assessor PIN/ref.	Date

Section below only to be completed by the learner once the assessor decision has been made and feedback given			
I confirm that the work carried out was my own, and that I received feedback from the Assessor	Learner name (print)	Learner signature	Date

Worksheet 30: Check and diagnose suspension faults

Procedure

▶ Damper/shock absorber operation – the vehicle body should move down as you press on it, bounce back just past the start point and then return to the rest position.

▶ Suspension bush condition – simple levering if appropriate should not show up excessive movement, cracks or separation of rubber bushes.

▶ Trim height – this is available from data books as a measurement usually taken from the wheel centre to a point on the car above.

Symptom	Possible causes of faults	Suggested action
Excessive pitch or roll when driving	Dampers/shock absorbers worn	Replace in pairs
Car sits lopsided	Broken spring	Replace in pairs
Car sits lopsided	Leak in hydraulic suspension	Rectify by replacing unit or fitting new pipes
Knocking noises	Excessive free-play in a suspension joint	Renew
Excessive tyre wear	Steering/suspension geometry incorrect (may be due to accident damage)	Check and adjust or replace any 'bent' or out of true components

Job card

Technician/learner name & date	Make and model	VIN no.		Reg. no.	Job/task no.

Customer's instructions/vehicle fault		Mileage			

Work carried out and recommendations (include PPE & special precautions taken)

Parts and labour	Price
Total	

Data and specifications used (include the actual figures)

Assessor report

	Assessment outcome	Passed (tick ✓)
1	The learner worked safely and minimised risks to themselves and others	
2	The learner correctly selected and used appropriate technical information	
3	The learner correctly selected and used appropriate tools and equipment	
4	The learner correctly carried out the task required using suitable methods and testing procedures	
5	The learner correctly recorded information and made suitable recommendations	

	Tick	Written feedback (with reference to assessment criteria) must be given when a learner is referred
Pass: I confirm that the learner's work was to an acceptable standard and met the assessment criteria of the unit		
Refer: The work carried out did not achieve the standards specified by the assessment criteria		

Assessor name (print)	Assessor PIN/ref.	Date

Section below only to be completed by the learner once the assessor decision has been made and feedback given

I confirm that the work carried out was my own, and that I received feedback from the Assessor	Learner name (print)	Learner signature	Date

Worksheet 31: Service manual steering system

Procedure

▶ Carry out basic checks as described in the 'Check steering components' worksheet.

▶ Check/top up steering gearbox oil (if appropriate).

▶ Check operation of column adjustment if fitted.

▶ Lubricate grease points on swivel joints/kingpins.

▶ Lubricate ball joints/track rod ends if appropriate (most are sealed for life).

▶ Road test.

Job card

Technician/learner name & date	Make and model	VIN no.	Reg. no.	Job/task no.

Customer's instructions/vehicle fault		Mileage		

Work carried out and recommendations (include PPE & special precautions taken)

Parts and labour	Price
Total	

Data and specifications used (include the actual figures)

Assessor report

	Assessment outcome	Passed (tick ✓)
1	The learner worked safely and minimised risks to themselves and others	
2	The learner correctly selected and used appropriate technical information	
3	The learner correctly selected and used appropriate tools and equipment	
4	The learner correctly carried out the task required using suitable methods and testing procedures	
5	The learner correctly recorded information and made suitable recommendations	

	Tick	Written feedback (with reference to assessment criteria) must be given when a learner is referred
Pass: I confirm that the learner's work was to an acceptable standard and met the assessment criteria of the unit		
Refer: The work carried out did not achieve the standards specified by the assessment criteria		

Assessor name (print)	Assessor PIN/ref.	Date

Section below only to be completed by the learner once the assessor decision has been made and feedback given			
I confirm that the work carried out was my own, and that I received feedback from the Assessor	Learner name (print)	Learner signature	Date

Worksheet 32: Check wheels and tyres for signs of damage and set tyre pressures

Procedure

▶ Jack and support the vehicle or raise on a wheel-free ramp.

▶ Check wheel for damage (each in turn including the spare).

▶ Check tyre for tread depth.

▶ Check tyre for damaged sidewalls.

▶ Check for signs of leakage from tyre and valve.

▶ Set pressures.

▶ Check tyre tread for feathering.

▶ Refit valve caps.

▶ Road test to check for wheel/tyre vibration, shimmy and noise.

▶ Rotate tyres according to manufacturer's recommendations.

▶ Report findings.

Job card

Technician/learner name & date	Make and model	VIN no.		Reg. no.	Job/task no.

Customer's instructions/vehicle fault		Mileage			

Work carried out and recommendations (include PPE & special precautions taken)

Parts and labour	Price
Total	

Data and specifications used (include the actual figures)

Assessor report

	Assessment outcome	Passed (tick ✓)
1	The learner worked safely and minimised risks to themselves and others	
2	The learner correctly selected and used appropriate technical information	
3	The learner correctly selected and used appropriate tools and equipment	
4	The learner correctly carried out the task required using suitable methods and testing procedures	
5	The learner correctly recorded information and made suitable recommendations	

	Tick	Written feedback (with reference to assessment criteria) must be given when a learner is referred
Pass: I confirm that the learner's work was to an acceptable standard and met the assessment criteria of the unit		
Refer: The work carried out did not achieve the standards specified by the assessment criteria		

Assessor name (print)	Assessor PIN/ref.	Date

Section below only to be completed by the learner once the assessor decision has been made and feedback given			
I confirm that the work carried out was my own, and that I received feedback from the Assessor	Learner name (print)	Learner signature	Date

Worksheet 33: Measure tyre tread and report on condition

Procedure

▶ Check: NSF OSF NSR OSR.

▶ Check and report minimum tread depth.

▶ Check and report condition.

▶ Check suitability for vehicle.

▶ Manufacturer's wheel nut torque.

▶ Set wheel nut torque.

▶ Manufacturer's pressure.

▶ Check/adjust pressure.

▶ Report to customer.

Job card

Technician/learner name & date	Make and model	VIN no.	Reg. no.	Job/task no.

Customer's instructions/vehicle fault	Mileage	

Work carried out and recommendations (include PPE & special precautions taken)

Parts and labour	Price
Total	

Data and specifications used (include the actual figures)

Assessor report

	Assessment outcome	Passed (tick ✓)
1	The learner worked safely and minimised risks to themselves and others	
2	The learner correctly selected and used appropriate technical information	
3	The learner correctly selected and used appropriate tools and equipment	
4	The learner correctly carried out the task required using suitable methods and testing procedures	
5	The learner correctly recorded information and made suitable recommendations	

	Tick	Written feedback (with reference to assessment criteria) must be given when a learner is referred
Pass: I confirm that the learner's work was to an acceptable standard and met the assessment criteria of the unit		
Refer: The work carried out did not achieve the standards specified by the assessment criteria		

Assessor name (print)	Assessor PIN/ref.	Date

Section below only to be completed by the learner once the assessor decision has been made and feedback given		
I confirm that the work carried out was my own, and that I received feedback from the Assessor	**Learner name (print)**	**Learner signature**	**Date**

Worksheet 34: Check wheels, tyres, trims and torque

Procedure

▶ Check carefully all around the tyre, inside and out.

▶ Look for tread wear and damage to the sidewalls.

▶ Feel the pattern of the tyre from left to right, and right to left, to check for signs of feathering.

▶ Check that the tyre shape is uniform and that there are no cuts or bulges.

▶ Check wheels. Alloys with signs of cracking should be replaced immediately.

▶ Make sure trims are fitted correctly and that the valve is positioned so it is not under stress from the trim.

▶ Wheel nuts/bolts should be tightened with a torque wrench.

▶ Check valves for signs of leakage and make sure the dust cap is fitted.

▶ Rock the valve gently from side to side to check for leaks.

▶ Check and adjust tyre air pressures.

▶ Measure wheel/tyre, axle and hub run out.

Job card

Technician/learner name & date	Make and model	VIN no.		Reg. no.	Job/task no.

Customer's instructions/vehicle fault		Mileage			

Work carried out and recommendations (include PPE & special precautions taken)

Parts and labour	Price
Total	

Data and specifications used (include the actual figures)

Assessor report

	Assessment outcome	Passed (tick ✓)
1	The learner worked safely and minimised risks to themselves and others	
2	The learner correctly selected and used appropriate technical information	
3	The learner correctly selected and used appropriate tools and equipment	
4	The learner correctly carried out the task required using suitable methods and testing procedures	
5	The learner correctly recorded information and made suitable recommendations	

	Tick	Written feedback (with reference to assessment criteria) must be given when a learner is referred
Pass: I confirm that the learner's work was to an acceptable standard and met the assessment criteria of the unit		
Refer: The work carried out did not achieve the standards specified by the assessment criteria		

Assessor name (print)	Assessor PIN/ref.	Date

Section below only to be completed by the learner once the assessor decision has been made and feedback given			
I confirm that the work carried out was my own, and that I received feedback from the Assessor	Learner name (print)	Learner signature	Date

Worksheet 35: Remove and refit front suspension strut and spring

Procedure

▶ Removal: apply the handbrake, jack up the front of the car and support on stands. Remove the road wheel. To prevent the lower arm assembly hanging down whilst the strut is removed, screw a wheel bolt into the hub, then wrap a piece of wire around the bolt and tie it to the car body. This will support the weight of the hub assembly. Unclip the brake hose and wiring harness from any clips on the base of the strut. Slacken and remove the lower bolts securing the suspension strut to the steering knuckle. From within the engine compartment, unscrew the strut upper mounting nuts and then carefully lower the strut assembly out from underneath the wing.

▶ Overhaul: warning – before dismantling the front suspension strut, a special tool to hold the coil spring in compression must be obtained. Any attempt to dismantle the strut without such a tool is likely to result in damage and/or injury. With the strut removed from the car, clean away all external dirt and then mount it upright in a vice. Fit the spring compressor, and compress the coil spring until all tension is relieved from the upper spring seat. Remove the cap from the top of the strut to gain access to the strut upper mounting retaining nut. Slacken the nut whilst holding the strut piston.

▶ Remove the mounting nut and washer, and lift off the rubber mounting plate. Remove the gasket and dished washer followed by the upper spring plate and upper spring seat. Lift off the coil spring and remove the lower spring seat. Examine all the components for wear, damage or deformation, and check the upper mounting bearing for smoothness of operation. Renew as necessary.

▶ Examine the strut for signs of fluid leakage. Check the strut piston for signs of pitting along its entire length, and check the strut body for signs of damage. While holding it in an upright position, test the operation of the strut by moving the piston through a full stroke, and then through short strokes (50 to 100 mm). In both cases, the resistance felt should be smooth and continuous. Renew if the resistance is jerky, uneven or if there is any visible sign of wear or damage. If any doubt exists about the condition of the coil spring, carefully remove the spring compressors, and check the spring for distortion and signs of cracking. Renew the spring if it is damaged.

▶ Fit the spring seat and coil spring onto the strut, making sure the spring end is correctly located against the strut stop. Fit the upper spring plate, aligning its stop with that of the seat, and fit the dished washer and gasket followed by the upper mounting plate. Locate the washer on the strut piston, then fit the mounting plate nut and tighten it to the specified torque. Refit all in reverse order.

Job card

Technician/learner name & date	Make and model	VIN no.		Reg. no.	Job/task no.

Customer's instructions/vehicle fault		Mileage		

Work carried out and recommendations (include PPE & special precautions taken)

Parts and labour	Price
Total	

Data and specifications used (include the actual figures)

Assessor report

	Assessment outcome	Passed (tick ✓)
1	The learner worked safely and minimised risks to themselves and others	
2	The learner correctly selected and used appropriate technical information	
3	The learner correctly selected and used appropriate tools and equipment	
4	The learner correctly carried out the task required using suitable methods and testing procedures	
5	The learner correctly recorded information and made suitable recommendations	

	Tick	Written feedback (with reference to assessment criteria) must be given when a learner is referred
Pass: I confirm that the learner's work was to an acceptable standard and met the assessment criteria of the unit		
Refer: The work carried out did not achieve the standards specified by the assessment criteria		

Assessor name (print)	Assessor PIN/ref.	Date

Section below only to be completed by the learner once the assessor decision has been made and feedback given			
I confirm that the work carried out was my own, and that I received feedback from the Assessor	Learner name (print)	Learner signature	Date

Worksheet 36: Remove, inspect and replace tie rods

Procedure

▶ Remove:

- Disconnect the battery, apply the handbrake and slacken the road wheel nuts. Raise the front of the vehicle, support it on axle stands and remove the front wheels.

- Remove the bolts securing the tie rod to the lower suspension arm.

- Disconnect stabilizer bar connecting link (if fitted). Remove the forward nut, retainer and insulator on the tie rod.

- Remove the tie rod complete with spacer, rear insulator and rear lock nut.

- Remove spacer, rear insulator and rear lock nut from the tie rod.

▶ Inspect:

- Inspect the tie bar for bending and condition of threads. Inspect condition of bushes, replace if needed.

▶ Refit:

- Assemble rear nut, retainer, spacer and rear insulator to tie rod. Relocate tie rod through frame (ensure spacer tube is located inside rear insulator).

- Replace front insulator, retainer and nut to front end of tie rod.

- Reconnect stabilizer bar connecting link (if fitted).

- Replace the bolts securing the tie rod to the lower suspension arm and tighten to correct torque.

- Replace road wheel, remove axle stands and lower vehicle to the ground. Fully tighten road wheel to the correct torque.

▶ Check castor and wheel alignment. Reset if required.

Note: wheel alignment must be measured and reset after castor has been adjusted.

Job card

Technician/learner name & date	Make and model	VIN no.		Reg. no.	Job/task no.

Customer's instructions/vehicle fault		Mileage	

Work carried out and recommendations (include PPE & special precautions taken)

	Price
Parts and labour	
Total	

Data and specifications used (include the actual figures)

Assessor report

	Assessment outcome	*Passed (tick ✓)*
1	The learner worked safely and minimised risks to themselves and others	
2	The learner correctly selected and used appropriate technical information	
3	The learner correctly selected and used appropriate tools and equipment	
4	The learner correctly carried out the task required using suitable methods and testing procedures	
5	The learner correctly recorded information and made suitable recommendations	

	Tick	Written feedback (with reference to assessment criteria) must be given when a learner is referred
Pass: I confirm that the learner's work was to an acceptable standard and met the assessment criteria of the unit		
Refer: The work carried out did not achieve the standards specified by the assessment criteria		

Assessor name (print)	Assessor PIN/ref.	Date

Section below only to be completed by the learner once the assessor decision has been made and feedback given			
I confirm that the work carried out was my own, and that I received feedback from the Assessor	Learner name (print)	Learner signature	Date

Worksheet 37: Service disc brakes

Procedure

▶ Jack up and support the vehicle on stands or use a suitable hoist. Remove the appropriate wheels.

▶ Inspect the brake pads. Recommendations vary slightly but in most cases the pads should be replaced if the lining is less than 1.5 mm. Methods of pad removal vary so check the manufacturer's data. However, most types are quite simple. The method described here relates to the type where part of the caliper is removed.

▶ Turn the steering to a lock position to allow easy access to the caliper and pads. Wash the caliper and pad assembly using a proprietary brake cleaner or suitable extractor.

▶ If necessary, remove some brake fluid from the reservoir. This is because when the piston is pushed back to allow new pads to be fitted, fluid can overflow. If a retaining bolt clip is fitted, it should be removed. Undo both caliper piston fixing bolts. Many types require an Allen key.

▶ Rock the assembly side to side. This moves the pads and pushes the piston in, just far enough to allow the caliper piston to be removed. Withdraw the pads. Use a small lever to help if a spring clip holds one of the pads into the piston. Keep the pads to show to the customer if necessary and then dispose of them in line with current regulations. Examine the disc for grooves and corrosion.

▶ Use a G/C clamp to push the caliper piston fully home. Fit the new pads in position together with anti-squeal shims if fitted. Some manufacturers recommend that copper grease be applied to the back and sides of each pad. Take care not to contaminate the lining. Repeat the process on the other side of the vehicle. Pads on both sides must always be replaced as a set.

▶ Refit the caliper and tighten all bolts to the recommended torque. Pump the brake pedal until it feels hard. This is to make sure the pads are moved fully into position. Double check correct fitment and then refit the road wheels and tighten wheel nuts to recommended torque. Lower the vehicle to the ground.

Job card

Technician/learner name & date	Make and model	VIN no.		Reg. no.	Job/task no.

Customer's instructions/vehicle fault		Mileage	

Work carried out and recommendations (include PPE & special precautions taken)

Parts and labour	Price
Total	

Data and specifications used (include the actual figures)

Assessor report

	Assessment outcome	Passed (tick ✓)
1	The learner worked safely and minimised risks to themselves and others	
2	The learner correctly selected and used appropriate technical information	
3	The learner correctly selected and used appropriate tools and equipment	
4	The learner correctly carried out the task required using suitable methods and testing procedures	
5	The learner correctly recorded information and made suitable recommendations	

	Tick	Written feedback (with reference to assessment criteria) must be given when a learner is referred
Pass: I confirm that the learner's work was to an acceptable standard and met the assessment criteria of the unit		
Refer: The work carried out did not achieve the standards specified by the assessment criteria		

Assessor name (print)	Assessor PIN/ref.	Date

Section below only to be completed by the learner once the assessor decision has been made and feedback given			
I confirm that the work carried out was my own, and that I received feedback from the Assessor	Learner name (print)	Learner signature	Date

Worksheet 38: Inspect and measure brake disc thickness and run out

Procedure

▶ Jack up and support the vehicle. Remove the appropriate wheels. Select neutral if the wheels are on the driven axle.

▶ Lever the pads back just enough to allow the disc to rotate freely.

▶ Inspect the surface of the disc for signs of cracking and grooves. Small grooves are to be expected after a period of use. Grooves deeper than about 0.4 mm are usually considered excessive.

▶ Using a micrometer, measure the thickness of the disc at several different places around the disc, towards the centre and towards the outer edge.

▶ Compare the readings to the manufacturer's specifications. Some manufacturers stamp the minimum thickness just inside the centre of the disc.

▶ Mount a dial gauge (dial indicator) on a magnetic, or other appropriate type of stand, with the plunger running about 15 mm in from the outer edge of the disc.

▶ Zero the gauge and rotate the disc. Take note of changes in the dial gauge reading.

▶ Refinish a grooved disc if allowable. Consult manufacturer's recommendations.

▶ Refer to the manufacturer's specifications for maximum allowable run out. As a guide, 0.15 mm is usually considered the limit. Replace discs with excessive run out.

▶ Refit all components and lower the vehicle to the ground.

Job card

Technician/learner name & date	Make and model	VIN no.		Reg. no.	Job/task no.

Customer's instructions/vehicle fault			Mileage		

Work carried out and recommendations (include PPE & special precautions taken)

Parts and labour	Price
Total	

Data and specifications used (include the actual figures)

Assessor report

	Assessment outcome	Passed (tick ✓)
1	The learner worked safely and minimised risks to themselves and others	
2	The learner correctly selected and used appropriate technical information	
3	The learner correctly selected and used appropriate tools and equipment	
4	The learner correctly carried out the task required using suitable methods and testing procedures	
5	The learner correctly recorded information and made suitable recommendations	

	Tick	Written feedback (with reference to assessment criteria) must be given when a learner is referred
Pass: I confirm that the learner's work was to an acceptable standard and met the assessment criteria of the unit		
Refer: The work carried out did not achieve the standards specified by the assessment criteria		

Assessor name (print)	Assessor PIN/ref.	Date

Section below only to be completed by the learner once the assessor decision has been made and feedback given			
I confirm that the work carried out was my own, and that I received feedback from the Assessor	Learner name (print)	Learner signature	Date

Transmission

Worksheet 39: Service transmission system

Procedure

Important note: always check manufacturer's data for the correct lubricant type and if transmission drained and refilled or just checked.

▶ Jack up and support the vehicle or raise it on a hoist.

▶ Check transmission oil level:

- Remove the plug in the side of the gearbox. Note that two plugs are fitted, one to drain the oil at the bottom and the top one for filling up. If the oil is not at the bottom of the opening, use a pump or 'squeeze' bottle to top up until the oil just runs out of the hole.

- Refit the plug.

- If appropriate, repeat the previous procedure for the overdrive unit and final drive assembly.

▶ Drain and refill:

- Drain and refill transmission oil. Note this is best done after the transmission is warm i.e. after a road test.

- Inspect transmission/transaxle case, extension housing, case mounting surfaces and vents.

- Clean the areas around the plugs and any seals by wiping with a cloth.

- Road test the vehicle and then check for leaks.

- Double check that any plugs and covers, which were removed, have been refitted securely. Lower the vehicle to the ground.

Job card

Technician/learner name & date	Make and model	VIN no.	Reg. no.	Job/task no.

Customer's instructions/vehicle fault	Mileage	

Work carried out and recommendations (include PPE & special precautions taken)

Parts and labour	Price
Total	

Data and specifications used (include the actual figures)

Assessor report

	Assessment outcome	Passed (tick ✓)
1	The learner worked safely and minimised risks to themselves and others	
2	The learner correctly selected and used appropriate technical information	
3	The learner correctly selected and used appropriate tools and equipment	
4	The learner correctly carried out the task required using suitable methods and testing procedures	
5	The learner correctly recorded information and made suitable recommendations	

	Tick	Written feedback (with reference to assessment criteria) must be given when a learner is referred
Pass: I confirm that the learner's work was to an acceptable standard and met the assessment criteria of the unit		
Refer: The work carried out did not achieve the standards specified by the assessment criteria		

Assessor name (print)	Assessor PIN/ref.	Date

Section below only to be completed by the learner once the assessor decision has been made and feedback given			
I confirm that the work carried out was my own, and that I received feedback from the Assessor	Learner name (print)	Learner signature	Date

Worksheet 40: Service automatic transmission system

Procedure

▶ Fit appropriate car protection kit.

▶ Warm up the transmission, by road test if possible. Raise the vehicle on a suitable hoist.

▶ Loosen and remove the drain plug if fitted and drain the oil into a suitable catch pan. If a drain plug is not fitted, undo the sump/pan bolts and drain carefully. Take care with hot oil; make sure you wear suitable protection such as rubber gloves and overalls. Drain the torque converter if a plug is fitted.

▶ Remove the sump/pan completely.

▶ Remove the transmission fluid filter/screen. This should be cleaned and replaced or renewed.

▶ Inspect the sump/pan for any unusual residues. This is useful to help diagnose faults.

▶ Clean the sump/pan thoroughly inside and out. Make sure all traces of the old gasket are removed. Fit new gaskets and any sealing rings as required, and refit the sump/pan.

▶ Tighten all the bolts evenly to the specified torque setting. Refit drain plug if removed.

▶ Check the manufacturer's data for the correct oil and refill with the appropriate quantity.

Note: only use the recommended oil. Other types may cause serious damage.

▶ Apply parking/handbrake, start the engine and move the lever through all positions. Recheck the fluid level. Note that some manufacturers recommend the level to be checked with the engine running.

▶ Conduct a road test to check for correct operation, noise and vibration. Return to the workshop and double check for leaks. Remove protection kit and report your findings.

Job card

Technician/learner name & date	Make and model	VIN no.	Reg. no.	Job/task no.

Customer's instructions/vehicle fault	Mileage	

Work carried out and recommendations (include PPE & special precautions taken)

Parts and labour	Price
Total	

Data and specifications used (include the actual figures)

Assessor report

	Assessment outcome	Passed (tick ✓)
1	The learner worked safely and minimised risks to themselves and others	
2	The learner correctly selected and used appropriate technical information	
3	The learner correctly selected and used appropriate tools and equipment	
4	The learner correctly carried out the task required using suitable methods and testing procedures	
5	The learner correctly recorded information and made suitable recommendations	

	Tick	Written feedback (with reference to assessment criteria) must be given when a learner is referred
Pass: I confirm that the learner's work was to an acceptable standard and met the assessment criteria of the unit		
Refer: The work carried out did not achieve the standards specified by the assessment criteria		

Assessor name (print)	Assessor PIN/ref.	Date

Section below only to be completed by the learner once the assessor decision has been made and feedback given

I confirm that the work carried out was my own, and that I received feedback from the Assessor	Learner name (print)	Learner signature	Date

Worksheet 41: Check driveline components

Procedure

▶ RWD and 4WD systems:

- Select neutral position and raise the vehicle on a hoist. Check for oil leaks from the propshaft sliding joint at the gearbox output and from the differential/final drive input. Check for leaks from the universal joints.

- Hold one part of each UJ in turn and rotate the other part back and forth. Check for freeplay.

- Check the security of all fixing bolts (usually four at each end).

- Check that any balance weights on the shaft are secure.

- If a centre bearing is fitted, check the bearing by rotating the propshaft (free the wheels for this). Make sure the mountings are secure and that the rubber is in good condition.

- If a donut drive is fitted, check it for security and condition.

▶ FWD and 4WD systems:

- Check for leaks from the differential output seals. Check for leaks from the driveshaft boots/gaiters. Make sure the clips hold the boots securely. Check the security of all fixing bolts (used on the inner part of some driveshafts).

- The driveshaft should have some axial (back and forth lengthways) movement. However, there should be little or no rotational movement in the CV joints. Check this by holding either side of the joints and rocking.

- Check that any balance weights on the shaft are secure. If fitted, check that the damper is secure.

- Check the driveshaft main nut for security and that the locking method is intact (split pin or lock tab usually).

Job card

Technician/learner name & date	Make and model	VIN no.		Reg. no.	Job/task no.

Customer's instructions/vehicle fault		Mileage			

Work carried out and recommendations (include PPE & special precautions taken)

Parts and labour	Price
Total	

Data and specifications used (include the actual figures)

Assessor report

Assessment outcome		Passed (tick ✓)
1	The learner worked safely and minimised risks to themselves and others	
2	The learner correctly selected and used appropriate technical information	
3	The learner correctly selected and used appropriate tools and equipment	
4	The learner correctly carried out the task required using suitable methods and testing procedures	
5	The learner correctly recorded information and made suitable recommendations	

	Tick	Written feedback (with reference to assessment criteria) must be given when a learner is referred
Pass: I confirm that the learner's work was to an acceptable standard and met the assessment criteria of the unit		
Refer: The work carried out did not achieve the standards specified by the assessment criteria		

Assessor name (print)	Assessor PIN/ref.	Date

Section below only to be completed by the learner once the assessor decision has been made and feedback given			
I confirm that the work carried out was my own, and that I received feedback from the Assessor	Learner name (print)	Learner signature	Date

Worksheet 42: Service rear wheel drive propshaft

Procedure

▶ Apply the handbrake and raise the vehicle on a hoist.

▶ Check propshaft for security and signs of damage.

▶ Check that any balance weights are secure.

▶ Check for oil leaks from the gearbox output seal where the propshaft sliding joint fits.

▶ Check universal joints for signs of leakage (grease).

▶ Using a grease gun, pump new grease into each universal joint (if possible).

▶ Clean off excessive grease.

▶ Check all mounting bolts for security.

Job card

Technician/learner name & date	Make and model	VIN no.		Reg. no.	Job/task no.

Customer's instructions/vehicle fault		Mileage	

Work carried out and recommendations (include PPE & special precautions taken)

	Price
Parts and labour	
Total	

Data and specifications used (include the actual figures)

Assessor report

Assessment outcome		Passed (tick ✓)
1	The learner worked safely and minimised risks to themselves and others	
2	The learner correctly selected and used appropriate technical information	
3	The learner correctly selected and used appropriate tools and equipment	
4	The learner correctly carried out the task required using suitable methods and testing procedures	
5	The learner correctly recorded information and made suitable recommendations	

	Tick	Written feedback (with reference to assessment criteria) must be given when a learner is referred
Pass: I confirm that the learner's work was to an acceptable standard and met the assessment criteria of the unit		
Refer: The work carried out did not achieve the standards specified by the assessment criteria		

Assessor name (print)	Assessor PIN/ref.	Date

Section below only to be completed by the learner once the assessor decision has been made and feedback given			
I confirm that the work carried out was my own, and that I received feedback from the Assessor	Learner name (print)	Learner signature	Date

Worksheet 43: Service final drive and differential

Procedure

▶ Jack up and support the vehicle or raise it on a hoist.

▶ Inspect the area around the final drive and differential unit for oil leaks. If necessary clean off old oil, road test and check again. Pay particular attention to the main gasket seals and the driveshaft output oil seals and/or the pinion input seal.

▶ Remove the filler/level plug and check the oil level. The oil should be level with or just below the threads of the plug. Check with a finger or probe if necessary.

▶ If topping off is necessary, refer to the manufacturer's specifications for the correct oil. On most front wheel drive cars, the oil for the final drive and differential is the same as for the main gearbox because the units are combined.

▶ Some vehicles should have the oil changed at certain intervals. If this is the case, drain out the old oil into a tray. It is better to do this after a road test during which time the oil will become warmer and therefore drain out more easily. Some rear axle final drive and differential units do not have a drain plug. In this case, the cover must be removed to drain oil.

▶ On rear wheel drive vehicles with fixed axles and halfshafts, it may be necessary to check for oil leaks into the brake drums on the rear. This would normally be carried out during servicing of the brakes.

▶ Refit any plugs and covers that were removed. Lower the vehicle to the ground.

Job card

Technician/learner name & date	Make and model	VIN no.	Reg. no.	Job/task no.

Customer's instructions/vehicle fault	Mileage

Work carried out and recommendations (include PPE & special precautions taken)

Parts and labour	Price
Total	

Data and specifications used (include the actual figures)

Assessor report

Assessment outcome	Passed (tick ✓)
1	The learner worked safely and minimised risks to themselves and others	
2	The learner correctly selected and used appropriate technical information	
3	The learner correctly selected and used appropriate tools and equipment	
4	The learner correctly carried out the task required using suitable methods and testing procedures	
5	The learner correctly recorded information and made suitable recommendations	

	Tick	Written feedback (with reference to assessment criteria) must be given when a learner is referred
Pass: I confirm that the learner's work was to an acceptable standard and met the assessment criteria of the unit		
Refer: The work carried out did not achieve the standards specified by the assessment criteria		

Assessor name (print)	Assessor PIN/ref.	Date

Section below only to be completed by the learner once the assessor decision has been made and feedback given			
I confirm that the work carried out was my own, and that I received feedback from the Assessor	Learner name (print)	Learner signature	Date

Electrical

Worksheet 44: Service electrical system

Procedure

▶ Battery:

- If the terminals require cleaning, fit a memory keeper and disconnect the earth/ground lead.

- Clean the battery posts and terminals. Use a wire brush and hot water as required. If hot water is used, follow this with copious amounts of cold water to wash away any acid from the paintwork.

- Dry the terminals and posts. Apply battery grease or petroleum jelly.

- If the battery is not sealed, top up with deionized water to a few millimetres above the plates.

- Check the battery state of charge using a voltmeter or hydrometer. Recharge if necessary OFF the vehicle.

- Refit the battery and terminals, and ensure they are secure. Restore electronic memory functions if memory keeper not used.

▶ General:

- Check that spare fuses are fitted. Carry out a general inspection looking for loose connections and damaged wires.

▶ System operation:

- Run through *all* electrical systems in turn to check for correct operation.

- Make sure that simple items such as the washer fluid bottle are secure and topped off.

Job card

Technician/learner name & date	Make and model	VIN no.	Reg. no.	Job/task no.

Customer's instructions/vehicle fault		Mileage		

Work carried out and recommendations (include PPE & special precautions taken)

Parts and labour	Price
Total	

Data and specifications used (include the actual figures)

Assessor report

	Assessment outcome	Passed (tick ✓)
1	The learner worked safely and minimised risks to themselves and others	
2	The learner correctly selected and used appropriate technical information	
3	The learner correctly selected and used appropriate tools and equipment	
4	The learner correctly carried out the task required using suitable methods and testing procedures	
5	The learner correctly recorded information and made suitable recommendations	

	Tick	Written feedback (with reference to assessment criteria) must be given when a learner is referred
Pass: I confirm that the learner's work was to an acceptable standard and met the assessment criteria of the unit		
Refer: The work carried out did not achieve the standards specified by the assessment criteria		

Assessor name (print)	Assessor PIN/ref.	Date

Section below only to be completed by the learner once the assessor decision has been made and feedback given

I confirm that the work carried out was my own, and that I received feedback from the Assessor	Learner name (print)	Learner signature	Date

Worksheet 45: Remove and refit electrical components

Procedure

▶ This worksheet is generic and can be applied to many systems. However, refer to manufacturer's procedures for specific information.

▶ Fit a memory keeper to prevent changing stored settings in electronic control units and in car entertainment systems.

▶ Remove the battery earth/ground lead.

▶ If removing the battery, disconnect the supply connection, remove the casing clamp and remove the battery.

▶ Take care to keep it level so that no electrolyte (sulphuric acid) is spilt.

▶ For other components, disconnect their supply wires. Making a note or suitable sketch where necessary of the connections will save time when refitting.

▶ For some components, it may be necessary to remove other parts to allow easy access. For example, an exhaust shield may need to be removed before the alternator can be disconnected.

▶ Disconnect linkages and/or peripherals.

▶ Undo all mountings and remove the unit from the vehicle.

▶ Refitting is a reversal of the removal process.

▶ The last job is always to reconnect the battery earth/ground lead.

Job card

Technician/learner name & date	Make and model	VIN no.		Reg. no.	Job/task no.

Customer's instructions/vehicle fault		Mileage			

Work carried out and recommendations (include PPE & special precautions taken)

Parts and labour	Price
Total	

Data and specifications used (include the actual figures)

Assessor report

Assessment outcome		Passed (tick ✓)
1	The learner worked safely and minimised risks to themselves and others	
2	The learner correctly selected and used appropriate technical information	
3	The learner correctly selected and used appropriate tools and equipment	
4	The learner correctly carried out the task required using suitable methods and testing procedures	
5	The learner correctly recorded information and made suitable recommendations	

	Tick	Written feedback (with reference to assessment criteria) must be given when a learner is referred
Pass: I confirm that the learner's work was to an acceptable standard and met the assessment criteria of the unit		
Refer: The work carried out did not achieve the standards specified by the assessment criteria		

Assessor name (print)	Assessor PIN/ref.	Date

Section below only to be completed by the learner once the assessor decision has been made and feedback given		
I confirm that the work carried out was my own, and that I received feedback from the Assessor	**Learner name (print)**	**Learner signature**	**Date**

Worksheet 46: Check all vehicle lights for correct operation

Procedure

▶ Note that if the left or right turn signals, or the hazards work, the unit is probably functioning correctly.

▶ Remember that a flasher unit is designed to flash at a different rate when a bulb is blown.

▶ Turn signal flasher units are usually located either as part of the fuse box or on the steering column.

▶ Remove covers or shrouds as necessary.

▶ Most types of flasher unit simply pull out of the socket.

▶ Replace by pushing the new unit into the holder.

▶ Make sure the new unit is the correct one for the vehicle. Note that more powerful units may be required if a towing socket is fitted.

▶ After renewal, make sure indicators and hazard lights operate.

Job card

Technician/learner name & date	Make and model	VIN no.	Reg. no.	Job/task no.

Customer's instructions/vehicle fault		Mileage		

Work carried out and recommendations (include PPE & special precautions taken)

Parts and labour	Price
Total	

Data and specifications used (include the actual figures)

Assessor report

	Assessment outcome	Passed (tick ✓)
1	The learner worked safely and minimised risks to themselves and others	
2	The learner correctly selected and used appropriate technical information	
3	The learner correctly selected and used appropriate tools and equipment	
4	The learner correctly carried out the task required using suitable methods and testing procedures	
5	The learner correctly recorded information and made suitable recommendations	

	Tick	Written feedback (with reference to assessment criteria) must be given when a learner is referred
Pass: I confirm that the learner's work was to an acceptable standard and met the assessment criteria of the unit		
Refer: The work carried out did not achieve the standards specified by the assessment criteria		

Assessor name (print)	Assessor PIN/ref.	Date

Section below only to be completed by the learner once the assessor decision has been made and feedback given			
I confirm that the work carried out was my own, and that I received feedback from the Assessor	Learner name (print)	Learner signature	Date

Worksheet 47: Check all vehicle lights for condition and security

Procedure

▶ Obtain technical/procedural data from manuals and information from vehicle owner/driver.

Light (lenses, reflectors)	Condition
Headlights	
Sidelights	
Rear/tail lights	
Stoplights	
Fog lights (front and rear)	
Number/license plate light	
Turn signals	
Spot/driving lights	

Job card

Technician/learner name & date	Make and model	VIN no.		Reg. no.	Job/task no.

Customer's instructions/vehicle fault	Mileage	

Work carried out and recommendations (include PPE & special precautions taken)

	Price
Parts and labour	
Total	

Data and specifications used (include the actual figures)

Assessor report

	Assessment outcome	Passed (tick ✓)
1	The learner worked safely and minimised risks to themselves and others	
2	The learner correctly selected and used appropriate technical information	
3	The learner correctly selected and used appropriate tools and equipment	
4	The learner correctly carried out the task required using suitable methods and testing procedures	
5	The learner correctly recorded information and made suitable recommendations	

	Tick	Written feedback (with reference to assessment criteria) must be given when a learner is referred
Pass: I confirm that the learner's work was to an acceptable standard and met the assessment criteria of the unit		
Refer: The work carried out did not achieve the standards specified by the assessment criteria		

Assessor name (print)	Assessor PIN/ref.	Date

Section below only to be completed by the learner once the assessor decision has been made and feedback given			
I confirm that the work carried out was my own, and that I received feedback from the Assessor	Learner name (print)	Learner signature	Date

Worksheet 48: Service/check seatbelt operation

Procedure

▶ Caution: some seatbelts incorporate tensioners that operate in the event of a collision. Do not attempt to dismantle these systems without reference to the manufacturer's data.

▶ Check for twisted webbing due to incorrect alignment. Adjust as required.

▶ Fully extend each belt in turn and inspect the webbing. Look for cuts, damage, broken threads, colour fading and bowed webbing. If any damage is noted, the belt should be renewed. Follow the manufacturer's procedures for this process.

▶ If the belt will not extend, check for contamination and twisting.

▶ Insert the tongue of each belt into its buckle. Pull hard to make sure it locks in place. Follow manufacturer's procedures to renew if in any doubt.

▶ Push the button to make sure the belt releases from the buckle easily.

▶ Pull each belt in turn fully out and make sure it retracts. It is acceptable to guide the belt home and prevent twisting. However, the spring should pull the belt into its fully retracted position.

▶ Pull sharply on each belt in turn to check that it locks up.

▶ Drive the vehicle in an area away from other traffic, and brake sharply from about 10 mph/16 kph. The driver's belt should lock and hold you in position. Use an assistant to check the other belts if necessary.

Job card

Technician/learner name & date	Make and model	VIN no.	Reg. no.	Job/task no.

Customer's instructions/vehicle fault		Mileage	

Work carried out and recommendations (include PPE & special precautions taken)

Parts and labour	Price

Total	

Data and specifications used (include the actual figures)

Assessor report

	Assessment outcome	Passed (tick ✓)
1	The learner worked safely and minimised risks to themselves and others	
2	The learner correctly selected and used appropriate technical information	
3	The learner correctly selected and used appropriate tools and equipment	
4	The learner correctly carried out the task required using suitable methods and testing procedures	
5	The learner correctly recorded information and made suitable recommendations	

	Tick	Written feedback (with reference to assessment criteria) must be given when a learner is referred
Pass: I confirm that the learner's work was to an acceptable standard and met the assessment criteria of the unit		
Refer: The work carried out did not achieve the standards specified by the assessment criteria		

Assessor name (print)	Assessor PIN/ref.	Date

Section below only to be completed by the learner once the assessor decision has been made and feedback given			
I confirm that the work carried out was my own, and that I received feedback from the Assessor	Learner name (print)	Learner signature	Date

Worksheet 49: Service body electrical system

Procedure

▶ This is a simple service operation but it is important that the tasks are carried out correctly.

▶ Check washer and wiper operation. Top up the washer fluid as required.

▶ Inspect the wiper blades and recommend renewal as appropriate.

▶ Adjust and clean washer jets as required. A pin is often useful for this.

▶ Check the operation of the audible warning device – the horn should make a noise when the button is pushed!

▶ Carry out a quick check of all body systems. Check windows, door locks and sunroof – for smooth operation (electric or manual).

▶ Disarm supplementary restraint system (SRS) air bag for duration of service. Rearm following service. Always follow manufacturer's safety procedures.

▶ Switch on the ignition and make sure all warning lights operate. Check in particular the SRS light, and that it goes out when the engine starts.

▶ Inspect all air bag positions for security.

▶ Use scan tool to check SRS system. Follow manufacturer's procedures.

▶ Check the in-car entertainment (ICE) system aerial for security.

▶ Test the ICE system for radio static, weak/intermittent/no radio reception, by listening to each speaker in turn. Make sure you reset the unit to the customer's preferences.

▶ Road test the vehicle and check the operation of the cruise control system.

Job card

Technician/learner name & date	Make and model	VIN no.	Reg. no.	Job/task no.

Customer's instructions/vehicle fault	Mileage	

Work carried out and recommendations (include PPE & special precautions taken)

Parts and labour	Price
Total	

Data and specifications used (include the actual figures)

Assessor report

	Assessment outcome	Passed (tick ✓)
1	The learner worked safely and minimised risks to themselves and others	
2	The learner correctly selected and used appropriate technical information	
3	The learner correctly selected and used appropriate tools and equipment	
4	The learner correctly carried out the task required using suitable methods and testing procedures	
5	The learner correctly recorded information and made suitable recommendations	

	Tick	Written feedback (with reference to assessment criteria) must be given when a learner is referred
Pass: I confirm that the learner's work was to an acceptable standard and met the assessment criteria of the unit		
Refer: The work carried out did not achieve the standards specified by the assessment criteria		

Assessor name (print)	Assessor PIN/ref.	Date

Section below only to be completed by the learner once the assessor decision has been made and feedback given			
I confirm that the work carried out was my own, and that I received feedback from the Assessor	Learner name (print)	Learner signature	Date

Worksheet 50: Remove and replace headlight unit

Procedure

▶ Fit memory keeper and disconnect the battery earth/ground lead.

▶ Remove covers as required and disconnect wires to lights.

▶ Remove grill and/or trim as necessary for access to light unit fixings.

▶ Undo bolts and/or clips and remove light unit from the car.

▶ Refitting is a reverse of the removal process.

▶ Note that some manufacturers require special contact grease to be applied to the terminals. This makes for a good electrical contact and keeps water out.

▶ Check and adjust alignment.

▶ Check all other lights.

Job card

Technician/learner name & date	Make and model	VIN no.		Reg. no.	Job/task no.

Customer's instructions/vehicle fault		Mileage	

Work carried out and recommendations (include PPE & special precautions taken)

Parts and labour	Price
Total	

Data and specifications used (include the actual figures)

Assessor report

	Assessment outcome	Passed (tick ✓)
1	The learner worked safely and minimised risks to themselves and others	
2	The learner correctly selected and used appropriate technical information	
3	The learner correctly selected and used appropriate tools and equipment	
4	The learner correctly carried out the task required using suitable methods and testing procedures	
5	The learner correctly recorded information and made suitable recommendations	

	Tick	Written feedback (with reference to assessment criteria) must be given when a learner is referred
Pass: I confirm that the learner's work was to an acceptable standard and met the assessment criteria of the unit		
Refer: The work carried out did not achieve the standards specified by the assessment criteria		

Assessor name (print)	Assessor PIN/ref.	Date

Section below only to be completed by the learner once the assessor decision has been made and feedback given			
I confirm that the work carried out was my own, and that I received feedback from the Assessor	Learner name (print)	Learner signature	Date

Worksheet 51: Remove and refit windscreen wiper motor

Procedure

Note: this is a generic procedure for a motor that can be accessed from the engine compartment; refer to the specific manufacturer's instructions. Switch off the ignition. Mark the position of the wiper blades with masking tape, and remove the wiper arms.

▶ Raise the bonnet and remove rubber strip and/or covers from the heating/ventilation system area.

▶ Remove wiper motor cover panels.

▶ Remove retaining screws as appropriate and remove the wiring harness plug from the motor.

▶ Unscrew the large nut on the wiper spindles.

▶ Slacken and remove the motor mounting bracket screws.

▶ Manoeuvre the motor and drive linkage out from its fittings, and remove from the vehicle.

▶ Undo the nut on the wiper spindle after marking the position of the crank arm. Unscrew the motor fixing bolts and remove the motor.

▶ Refitting is a reversal of the removal process. However, note the following points.

▶ Connect the motor to the harness and run it (without the linkage) until it stops in the 'park' position as normal. Disconnect from the wiring.

▶ Refit the crank and linkage exactly as it was removed.

▶ After refitting the motor and linkage, run the motor and make sure the movement is correct *before* refitting the arms and blades.

▶ Finally, fit the arms and blades, wet the screen and check for correct operation at all speeds and settings. Check that the blades park correctly.

Job card

Technician/learner name & date	Make and model	VIN no.		Reg. no.	Job/task no.
Customer's instructions/vehicle fault		Mileage			

Work carried out and recommendations (include PPE & special precautions taken)

Parts and labour	Price
Total	

Data and specifications used (include the actual figures)

Assessor report

	Assessment outcome	Passed (tick ✓)
1	The learner worked safely and minimised risks to themselves and others	
2	The learner correctly selected and used appropriate technical information	
3	The learner correctly selected and used appropriate tools and equipment	
4	The learner correctly carried out the task required using suitable methods and testing procedures	
5	The learner correctly recorded information and made suitable recommendations	

	Tick	Written feedback (with reference to assessment criteria) must be given when a learner is referred
Pass: I confirm that the learner's work was to an acceptable standard and met the assessment criteria of the unit		
Refer: The work carried out did not achieve the standards specified by the assessment criteria		

Assessor name (print)	Assessor PIN/ref.	Date

Section below only to be completed by the learner once the assessor decision has been made and feedback given			
I confirm that the work carried out was my own, and that I received feedback from the Assessor	Learner name (print)	Learner signature	Date

Worksheet 52: Checking operation of the main instruments and warning lights

Procedure

Note: this is a generic test routine; refer to manufacturer's procedures and circuits for specific details.

▶ Switch the ignition on but do not start the engine.

▶ Check that the rev-counter and speedometer both read zero (or less).

▶ Check that the oil pressure warning light (WL) and charge WL are on.

▶ Check that the fuel gauge reads appropriate to fuel in the tank.

▶ Check that the temperature gauge reads appropriate to the engine temperature (best to start from cold if possible).

▶ Check for neutral or park, and start the engine.

▶ Oil light and charge light should go out as engine revs – and stay out at idle speed.

▶ Check rev-counter reads correctly by comparing with a test meter reading.

▶ Check fuel gauge reading remains constant and that temperature gauge rises to 'normal' level.

▶ Road test to check speedometer operation.

Job card

Technician/learner name & date	Make and model	VIN no.		Reg. no.	Job/task no.

Customer's instructions/vehicle fault		Mileage		

Work carried out and recommendations (include PPE & special precautions taken)

Parts and labour	Price
Total	

Data and specifications used (include the actual figures)

Assessor report

Assessment outcome		Passed (tick ✓)
1	The learner worked safely and minimised risks to themselves and others	
2	The learner correctly selected and used appropriate technical information	
3	The learner correctly selected and used appropriate tools and equipment	
4	The learner correctly carried out the task required using suitable methods and testing procedures	
5	The learner correctly recorded information and made suitable recommendations	

	Tick	Written feedback (with reference to assessment criteria) must be given when a learner is referred
Pass: I confirm that the learner's work was to an acceptable standard and met the assessment criteria of the unit		
Refer: The work carried out did not achieve the standards specified by the assessment criteria		

Assessor name (print)	Assessor PIN/ref.	Date

Section below only to be completed by the learner once the assessor decision has been made and feedback given			
I confirm that the work carried out was my own, and that I received feedback from the Assessor	Learner name (print)	Learner signature	Date

Worksheet 53: Check operation of heating and ventilation system

Procedure

▶ Start the engine and run until it is warm (use extraction if indoors).

▶ Check that the booster fan runs at all speeds. Switch off air conditioning if fitted.

▶ Set the temperature control to cold and the fan speed to a medium setting.

▶ Run through all direction settings and check that *cool* air is supplied.

▶ Set the temperature control to hot and the fan speed to a medium setting.

▶ Run through all direction settings and check that *hot* air is supplied.

▶ Check that a range of temperatures can be selected and that external or recirculated air can be used.

▶ Make sure all ventilation grills open and allow directional control.

▶ Check heated rear screen operation.

▶ Check heated front screen operation (if fitted).

Job card

Technician/learner name & date	Make and model	VIN no.		Reg. no.	Job/task no.
Customer's instructions/vehicle fault		Mileage			

Work carried out and recommendations (include PPE & special precautions taken)

Parts and labour	Price
Total	

Data and specifications used (include the actual figures)

Assessor report

Assessment outcome		Passed (tick ✓)
1	The learner worked safely and minimised risks to themselves and others	
2	The learner correctly selected and used appropriate technical information	
3	The learner correctly selected and used appropriate tools and equipment	
4	The learner correctly carried out the task required using suitable methods and testing procedures	
5	The learner correctly recorded information and made suitable recommendations	

	Tick	Written feedback (with reference to assessment criteria) must be given when a learner is referred
Pass: I confirm that the learner's work was to an acceptable standard and met the assessment criteria of the unit		
Refer: The work carried out did not achieve the standards specified by the assessment criteria		

Assessor name (print)		Assessor PIN/ref.	Date

Section below only to be completed by the learner once the assessor decision has been made and feedback given			
I confirm that the work carried out was my own, and that I received feedback from the Assessor	Learner name (print)	Learner signature	Date

Worksheet 54: Check central door locking and alarm operation

Note: different systems operate in different ways so check specific data as necessary.

▶ Use a scan tool where appropriate to check for alarm/central locking stored fault codes.

▶ Close all the doors and operate the central locking from the driver's door lock using the key manually. All doors and the tailgate should lock. If a double locking system is fitted, turning the key again double locks all openings.

▶ Check manually that all doors and openings have locked.

▶ Repeat the above procedure using the remote key if available.

▶ Repeat again from the passenger's door lock.

▶ Open one of the windows and then fully lock the car.

▶ Reach inside the car. If a movement sensor is incorporated, the alarm will sound! If not, reach in and open the door from the inside. The alarm should now sound! Press the remote or use the key in the driver's door to reset.

▶ Close all windows and lock the car.

Job card

Technician/learner name & date	Make and model	VIN no.		Reg. no.	Job/task no.

Customer's instructions/vehicle fault		Mileage		

Work carried out and recommendations (include PPE & special precautions taken)

Parts and labour	Price
Total	

Data and specifications used (include the actual figures)

Assessor report

	Assessment outcome	Passed (tick ✓)
1	The learner worked safely and minimised risks to themselves and others	
2	The learner correctly selected and used appropriate technical information	
3	The learner correctly selected and used appropriate tools and equipment	
4	The learner correctly carried out the task required using suitable methods and testing procedures	
5	The learner correctly recorded information and made suitable recommendations	

	Tick	Written feedback (with reference to assessment criteria) must be given when a learner is referred
Pass: I confirm that the learner's work was to an acceptable standard and met the assessment criteria of the unit		
Refer: The work carried out did not achieve the standards specified by the assessment criteria		

Assessor name (print)	Assessor PIN/ref.	Date

Section below only to be completed by the learner once the assessor decision has been made and feedback given			
I confirm that the work carried out was my own, and that I received feedback from the Assessor	Learner name (print)	Learner signature	Date

Worksheet 55: Remove and replace pre-engaged starter motor

Procedure

▶ Check the condition of the battery and the operation of the starter motor before removal. Disconnect the battery earth or ground lead. Label and disconnect the cables on the solenoid – solenoid feed from the starter switch and LT ignition feed to bypass the ballast resistor. Disconnect the main supply feed cable to the starter at either end (most convenient). Remove any components that restrict removal of the starter motor. Undo the starter motor securing bolts. (Socket wrenches and extensions are useful to gain access between the starter motor body and the engine block). Carefully remove the starter motor. Inspect the drive pinion (gear) and the one-way clutch. Check the starter motor casing and fitting flange for damage. Check the starter ring gear on the flywheel through the hole where the starter motor fits. Turn the engine at least one full revolution.

▶ Check all gear teeth for chipped or worn down sections both on individual teeth and around the full circumference of the gear. Uneven wear in one place is a frequent defect. Look at the teeth for wear marks that show that the starter pinion has been fully in mesh. To strip the motor fit into a bench vice and secure. To remove the solenoid, undo the feed cable into the motor and pull back out of the way. Undo the retaining screws in the casing. Carefully pull out the solenoid disconnecting the plunger from the pinion gear engagement lever. To strip the motor, remove the through bolts and the rear cover and brushes. The brushes may be available as replacement parts. Follow the manufacturer's instructions for replacement. Clean and inspect the commutator, armature and spindle bearings. Pull the pinion casing, pinion, engagement lever and armature from the main casing and field coils. Follow the manufacturer's instructions for replacing the drive pinion and one-way clutch.

▶ Reassemble in reverse order. Carry out a bench test with the starter securely held in a bench vice. Connect a battery with jump leads to the starter. Negative cable to the battery negative and the starter casing. Positive lead *only* to the battery positive and then keep the lead clear until testing. *Avoid* any connect of the positive lead to the starter casing, vice or bench. Test the solenoid operation by touching the lead to the solenoid terminal, which should click, and the pinion move along the spindle to the engaged position. The spindle may slowly revolve on some motors.

▶ Test the motor operation by touching the lead to the motor terminal. The motor should run at full speed. Connect the lead to the solenoid input terminal. The solenoid should not operate and the motor should not run. Use a jump lead to connect a feed to the solenoid low current terminal. The solenoid should operate and the motor should run. Refit the starter motor to the engine and check the operation.

Job card

Technician/learner name & date	Make and model	VIN no.	Reg. no.	Job/task no.

Customer's instructions/vehicle fault	Mileage	

Work carried out and recommendations (include PPE & special precautions taken)

Parts and labour	Price
Total	

Data and specifications used (include the actual figures)

Assessor report

	Assessment outcome	Passed (tick ✓)
1	The learner worked safely and minimised risks to themselves and others	
2	The learner correctly selected and used appropriate technical information	
3	The learner correctly selected and used appropriate tools and equipment	
4	The learner correctly carried out the task required using suitable methods and testing procedures	
5	The learner correctly recorded information and made suitable recommendations	

	Tick	Written feedback (with reference to assessment criteria) must be given when a learner is referred
Pass: I confirm that the learner's work was to an acceptable standard and met the assessment criteria of the unit		
Refer: The work carried out did not achieve the standards specified by the assessment criteria		

Assessor name (print)	Assessor PIN/ref.	Date

Section below only to be completed by the learner once the assessor decision has been made and feedback given			
I confirm that the work carried out was my own, and that I received feedback from the Assessor	Learner name (print)	Learner signature	Date

Worksheet 56: Remove and replace alternator

Procedure

▶ Disconnect the battery earth or ground lead. Unplug the multi-socket on the rear of the alternator or label and disconnect the cables. Slacken the alternator securing bolts, including the drive belt adjuster strap bolts. Mark the direction of rotation on the belt. Slacken the drive belt tension and remove the belt. Inspect the drive belt and drive pulleys for signs of wear, damage and slipping (glazing on sides).

▶ Refit in the reverse order and adjust the drive belt tension. Run the engine and check that the generator/ignition warning light comes on and then goes out as the engine speed increases. Connect a digital voltmeter and clamp on ammeter and check that the alternator output is correct. Compare with manufacturer's data.

Job card

Technician/learner name & date	Make and model	VIN no.		Reg. no.	Job/task no.
Customer's instructions/vehicle fault		**Mileage**			

Work carried out and recommendations (include PPE & special precautions taken)

Parts and labour	Price
Total	

Data and specifications used (include the actual figures)

Assessor report

	Assessment outcome	Passed (tick ✓)
1	The learner worked safely and minimised risks to themselves and others	
2	The learner correctly selected and used appropriate technical information	
3	The learner correctly selected and used appropriate tools and equipment	
4	The learner correctly carried out the task required using suitable methods and testing procedures	
5	The learner correctly recorded information and made suitable recommendations	

	Tick	Written feedback (with reference to assessment criteria) must be given when a learner is referred
Pass: I confirm that the learner's work was to an acceptable standard and met the assessment criteria of the unit		
Refer: The work carried out did not achieve the standards specified by the assessment criteria		

Assessor name (print)	Assessor PIN/ref.	Date

Section below only to be completed by the learner once the assessor decision has been made and feedback given			
I confirm that the work carried out was my own, and that I received feedback from the Assessor	**Learner name (print)**	**Learner signature**	**Date**

Printed in the United States
by Baker & Taylor Publisher Services